Blind Maps and Blue Dots

The Blurring of the Producer–User Divide
in the Production of Visual Information

Joost Grootens

Lars Müller Publishers

Contents

Introduction

In the future TV will be so good that the printed word will function
as an art form only.
— David Byrne, "In the Future," 1984

This book explores what contemporary mapmaking practices can reveal about
the ever-evolving field of graphic design. It consists of a textual part and a visual
part. The series of visualizations represent and summarize the research through
imagery. In this sense, this PhD dissertation on mapmaking is itself a map: it embod-
ies both the itinerary of my investigation—the text—and a chart for the reader to
find her way—the visuals.

I write this introduction at the end of the research process. With the clarity of
hindsight I can now describe my path as a straight line. The journey, however,
was less straightforward. The starting point was the need to understand changes
that were happening in my own graphic design and mapmaking practice. The
digitization of tools to record, create, edit, produce and distribute visual information has triggered a process of transformation of cartography and graphic design
that I believed (and still believe) to be a fundamental change.

In the early stages of this research I tried to grasp the essence of this transformation. The crux for me was the disappearance of the distinction between the
different roles in the process of the creation, the editing, production and use of
visual information. Up until the digital age, these positions were defined by the
distinct tasks performed by specialists like writers, editors, illustrators, graphic
designers, type setters, lithographers and printers. The introduction of the computer as universal tool, and the interactive and coproduction formats advanced
by Web 2.0 technologies, have blurred the distinction between production and
use and between designer and user.

For a better understanding of the disappearance of the producer–user divide
it was useful to learn about the theory of post-representational cartography. Its
premise is that producing and using are not consecutive stages in the life of a
map, but rather are parallel processes. With the disappearance of the distinction
between producer and user it was useful to shift the focus from the product to
the processes of production and use. Insights derived from post-representational
cartography, experiences from my own practice, as well as theories related to
the specific case studies guided me in the investigation into three mapmaking
practices.

Intuitively, I sensed that studying the transformation of cartography could help
to better understand the significant changes happening in the broader field of
graphic design. Mapmaking is a more clearly distinguishable discipline compared

with the expanded ill-defined field of graphic design as it has developed in the digital age. To better understand the blurring of the producer–user divide, I focused my research on players who have entered the field of mapmaking since the democratization of its tools. I selected three mapmaking practices by amateurs and technology companies that appropriated technologies of mapping, designing and publishing, and investigated their tools, strategies and output.

From the steps listed above I have omitted choices that turned out to be side roads rather than junctions on the research path. Over time, certain concepts in the inquiry became less important and were discarded or replaced by more appropriate terms. In the early stages the research question was centered on key concepts like "rhetoric," "digital technology," and "tools." In a research process that builds on intuition, like this one, many ideas are discarded because they turn out to be unusable. The elements that remained were "graphic design" as the field of research, "the disappearing distinction between producers and users" as the development that that field is going through, "contemporary mapmaking practices" as research object, and "post-representational cartography" as research method. The elements were combined into the following research question: What can a post-representational reading of contemporary mapmaking practices reveal about the blurring of the producer–user divide in graphic design?

Chapter 1, Concepts and Methods, looks into the key concepts used in the research. It introduces the fields of research, theoretical framework, and the methods that are employed in the investigation. The chapter will also address the role practice plays in the research and the choice of the book as dissertation format.

In Chapter 2, Positioning, the research is situated within the ever-evolving field of graphic design. The chapter proposes a new model to better understand the transformation of the field. Building on experiences from my design and mapmaking practice I will draw parallels between graphic design and cartography. In this chapter the visual concept of the blind map is introduced to describe the permanent emergent status of graphic products.

Chapters 3, 4 and 5 contain case studies of contemporary mapmaking practices by technology companies and amateurs.

In Chapter 3, The Blue Dot, different modes of cartographic thinking are used to assess online mapping service Google Maps. The chapter discusses an ambiguous approach to mapmaking as a response to post-representational cartography. I will introduce the blue dot, a visual concept that marks the disappearance of the producer–user divide.

In Chapter 4, The Strava Global Heatmap, a map produced by a social network for athletes, is investigated from a technological, economic, social and cultural point of view.

Chapter 5, The Situation in Syria, looks into the practices of amateur conflict mapmakers. The chapter discusses the differences between practices of specialists and nonspecialists in terms of visual strategies, production and publishing.

Chapter 6, Conclusion, proposes a post-representational approach to graphic design. The chapter also addresses the need for alternative and additional languages in multidisciplinary discourse, in research in general, and in artistic research in particular.

Notes on the Visualizations

The issues investigated in the visualizations in this book are the relationship between graphic design and the technologies that produce visual information, the development of these technologies over time, and the impact of software and hardware, and the systems in which they operate, on the production of visual information.

In essence, the visualizations in this book do two things: dissect objects and systems, and show how these evolve over time. The visualizations form series in which different aspects are highlighted or compared. The six series run parallel to the chapters of the discursive part of the book.

The aspects that the visualizations address are diverse and vary in scale, amount and visibility. This has led to a diversity of visualization formats: timelines, illustrations, scans, diagrams and legends. In general, the horizontal direction indicates time in the visualizations. The vertical direction shows difference in position, either geographically or in terms of proximity.

1   **Concepts and Methods**

In this chapter I will present the main concepts used in the research. I will then briefly introduce the fields of cartography and graphic design and discuss the role of practice in the investigation. Then I will address the theoretical framework, and the method of "exhausting and scraping" employed in this research. I will conclude by elucidating the choice of the book as dissertation format.

Where the description of a route only allows one kind of use, a map allows multiple applications. In this chapter I intend to do the same, both providing a more precise context to the research and opening it up for others to use it.

Put otherwise, whereas the path described in the introduction focused on *research* as a verb, on the action of systematically investigating, what follows below will concentrate on *research* as a noun: a systematic investigation of a subject with the intention of gaining new insights that are justified and made public.

Three Key Concepts: Information, Production, Visualization

The subject of this book is the production of visual information. The research focuses on the disappearing boundary between producers and users in graphic design. This field is situated between, and occasionally encompasses, the conception and production of visual information and its multiplication through reproduction methods. Until recently, this mainly concerned printed matter. The digitization of tools and of production formats has resulted in a radical shift of the positions of producers and users of graphic design. It has become difficult to uphold a strict division between the two. I will elaborate on this transformation later. First, I will unpack the three main elements that constitute the subject: information, production and visualization.

I regard *information* as "interpreted data." Data are things known or assumed as facts, such as geographic coordinates, measurements or statistics, but that have no built-in visual form. Design can convert data into information through a process of editing, organizing and translating, just like a story does in which facts are turned into a narrative. Knowledge can be understood as the capacity to interpret information through experience and/or education. Design can facilitate the interpretation of information and therefore support the process of knowledge transfer.

Production of information I regard to be the process of collecting data and editing, organizing and transforming it into information. This includes representing, reproducing and distributing the information, up to accessing, interpreting and using it. Until the digital age, a variety of specialists, each with their own expertise, were involved in this information chain, that is, in the consecutive steps of the production process. Digital technologies incorporated some of the expert activities, like layout, typesetting and image lithography. At the same time, by making the tools accessible to everybody, digital technologies opened up production activities to people who until then were not involved in the (full) process of transforming data into information, such as journalists. Digital technology also opened up the activities of graphic designers, giving them direct access to production technologies, and at the same time giving them access to the tools to create content. Digital modes of production thus changed the information chain from a one-way series of consecutive, compartmentalized tasks done by a variety of experts into a multidirectional blur of roles and activities.

Graphic design studio SJG, Amsterdam

The relationship between the subject of the research, the transformation of graphic design, and the place where the research was conducted, a graphic design studio, is an ambiguous one. The research was developed in parallel with the regular graphic design activities of a design studio. The technologies that were investigated were in some cases the same hardware and software with which the research was done.

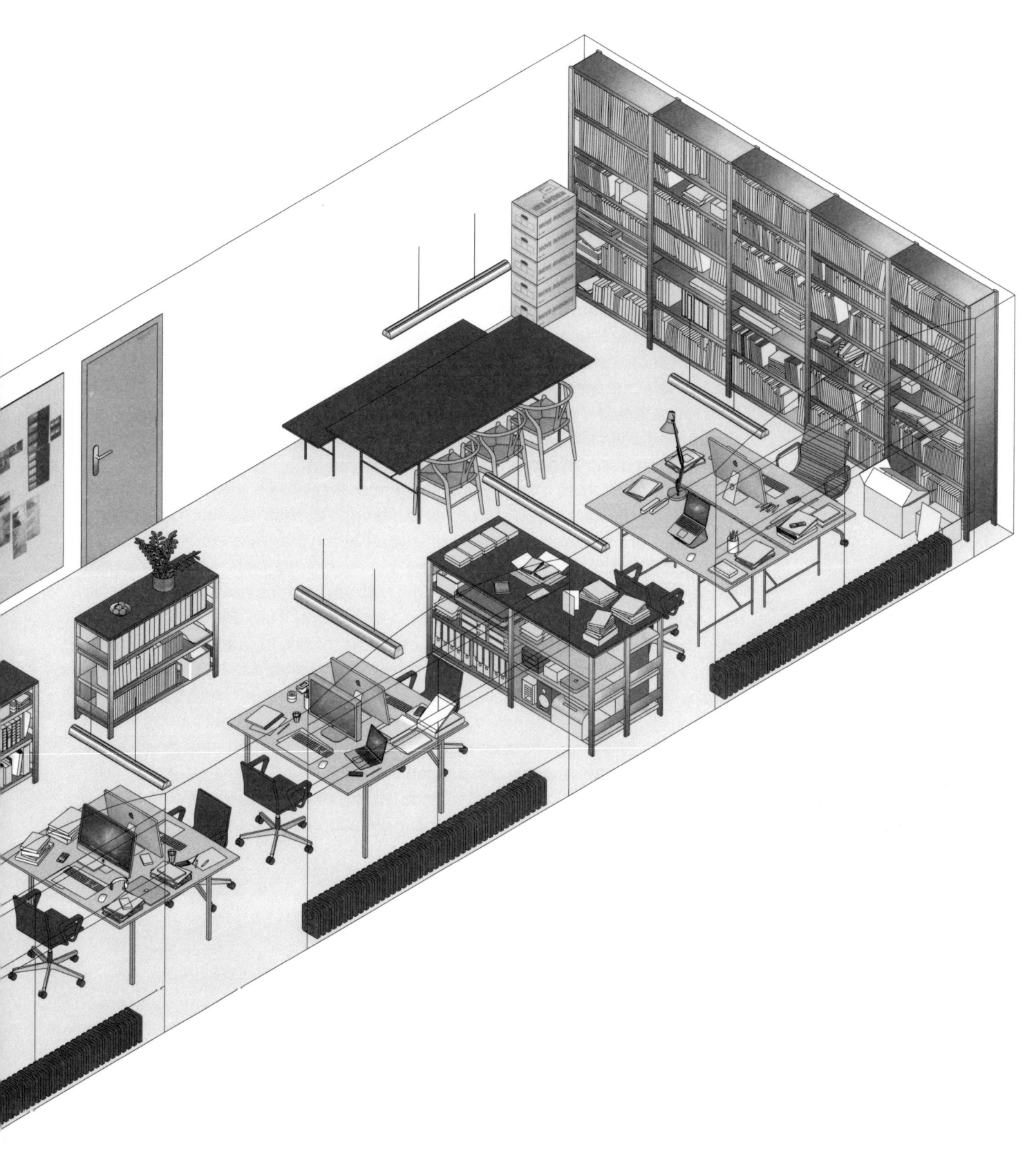

Put otherwise, graphic design originated in the late nineteenth century as a specialized activity concerned with creating, editing and reproducing visual information. Digitization of its means of production in the 1980s sparked a fundamental transformation of graphic design. The computer became a universal tool combining hitherto distinct specialized tasks. This, together with the interactive possibilities of digital media, and the easily accessible dissemination of visual information via networks such as the Internet, triggered the changes. As a result, the output of graphic design diversified enormously and now included new digital formats such as websites. It also included formats from other fields, such as animation. The strict division between the creators, editors, producers and users of visual information became blurred.

I will use the term *visualization* as specified in the *Oxford English Dictionary*'s first definition of the term: a "representation of an object, situation, or set of information as a chart or other image," and not in the sense of its second description, "the formation of a mental image of something."[1] For the verb *to visualize* I use the second definition in the *OED*, to "make something visible to the eye," rather than the first description, to "form a mental image of."[2]

This book refers to a variety of other concepts related to visuality and visualization. I use the term *visual strategies* to describe the use of visual means in the production of information to achieve an overall aim in the visualization. In Chapter 3, for instance, the term *visual strategies* is used to describe how the use of pale colors for the topographic layer in Google Maps de-emphasizes the geographic information, while the use of bright colors for names and pictograms in the same map draws attention to the addresses of places to visit. In Chapter 5 I make reference to "visibilization," a visualization that makes visible what is hidden, a concept by German cultural scientist Anne Huffschmid. In Chapter 4 I refer to American psychologist Michael Friendly's research into the history of information visualizations to address the shortcomings in the process of understanding new types of information visualizations. More in general, I discuss problems around visuality by citing Canadian philosopher of science Ian Hacking's ideas on the constructed nature of recorded images, and American philosopher John Dewey's "spectator theory of knowledge" (both in Chapter 5).

Cartography as Testing Ground for Graphic Design

The shift towards digital modes of production also affected other fields concerned with visual representation. Like graphic design, cartography (the study and practice of mapmaking) is shaped by technological developments of tools. Until the digital age, maps were produced by those in power, such as the state. Digital technology created easily accessible cartographic tools to collect data and to produce geographic information visualizations. Technology thus empowered new players to enter the field of cartography. With no prior knowledge of cartography, these new players started mapping different subjects in novel ways, occasionally resulting in new types of maps.

That is not to say that cartography and graphic design are the same, even though they share certain characteristics. The output of both disciplines is primarily visual, and both use similar tools to create, record, edit, produce and distribute.

But whereas cartography deals both with the study of maps, as well as with the collecting of geographic data and the transformation of that data into geographic information visualizations such as maps, graphic design does not concern the study of its own products as an end in itself. Furthermore, graphic design is not focused on a singular output but can manifest itself in a great variety of formats— posters, visual identities, wayfinding, advertising, packaging, books, magazines, signage, websites, animations, information visualizations, et cetera—in order to persuade and to inform. Although the distinction between the activities is less clear in the digital age, cartography, unlike graphic design, originally also included data collection. Graphic design, on the other hand, unlike cartography, originally also covered the preparation for production of its output.

Cartography is an interesting field of study when investigating the blurring of the producer–user divide in the production of visual information. The output of the mapmaking process is more singular than that of graphic design and therefore the effects of the digitization of the means of production are more clearly discernible. Furthermore, as mapmaking was a highly closed-off field dominated by powerful elites, the impact of digital technologies opening up cartography seems more considerable than the impact of similar developments on graphic design. A third reason for considering mapmaking as a testing ground for studying the blurring lines between producers and users is that this relation is more substantially theorized in cartography, for instance in post-representational cartography. Unlike other cartographic theories, post-representational cartography looks at the full process of mapmaking, from conception to use, from mapmaker to user. This processual approach to mapmaking regards the binary division between production and application, between producer and user, as no longer relevant. In my research, concepts and methods from this processual approach are used in three case studies of contemporary mapmaking practices of technology companies and amateurs.

Research Concerned with, and by Means of Practice

This research project originated from an urgency felt in both my graphic design practice and in my educational practice of teaching graphic designers. What are possible consequences of the radical transformation of graphic design, as described above, for my own practice and that of future designers? And in darker days: can I still morally justify teaching graphic designers when I am uncertain about the future of the field? The research in this book is practice-led, its aim is to acquire new insights into my practice. But the research is also practice-based in that it gains knowledge partly by means of practice and the outcomes of that practice.

There is a third way in which practice plays a role in this book. Besides research about and through practice, I will identify practices of nonspecialist mapmakers who entered the field following the democratization of the tools as described above. In addition to practice-led and practice-based, I will call this third strand practice-identified. I will use my own design processes, activities and output as a comparative praxis next to the ones I examine. Furthermore, to better understand maps of nonspecialist mapmakers, I will conceptualize alternative outcomes. In

1   Oxford English Dictionary, "visualization," accessed 24 September 2019, https://www.lexico.com/en/definition/visualization.

2   Oxford English Dictionary, "visualize," accessed 24 September 2019, https://www.lexico.com/en/definition/visualize.

A HP Laserjet black and white printer, A4 F Apple Display screen
B HP Color Laserjet printer, A3   G Apple MacBook Pro
C Epson scanner, A4   H Apple iPad
D Epson scanner, A3   I Apple iPhone
E Apple iMac   J Android phone

short, I will use my skills, awareness and methods as a designer and mapmaker to study mapmaking practices that formerly have not been labeled as such, and I will analyze and speculate about their output.

I anticipate that this "learning from," to quote Venturi, Scott Brown and Izenour's *Learning from Las Vegas* (1972), can help bridge the gap between the traditional field and the new players. The seminal book about the city in the Mojave Desert has been an important reference for me, because it was the ambition of the authors to try to make sense as architects of a "nonarchitecture." In doing so, they want to "withhold judgment," as the authors write in the introduction.[3] They use various formats to document their research, such as texts, photographs, drawings, maps and diagrams.

In writing this book I have aimed to address and question the role language plays in cross-disciplinary research. Jargon, the specific terminology used by specialists to describe a field or its processes, is an obstacle in conversations about graphic design beyond current disciplinary boundaries. An expanded understanding of graphic design will need to address its own characteristic modes of communication. I used my design practice to create a series of visualizations to describe and situate the practices that are investigated in this book. The visualizations are as much a method of doing research, gaining insights by studying a different overview of the content, as an exploration of alternative languages to represent the research.

Presenting different types of representation next to one another is a strategy I often employ in the design of books, especially in atlases. The method of showing a variety of modes of description, like maps, data visualizations, photography and text, questions each format individually while at the same time highlighting the ambiguous processes going on in representation. Avoiding a singular language liberates each particular representational format and allows each type of representation to be more outspoken. Individual representation types, like text or visualizations, do not have to individually carry the weight of being unequivocal or complete as the totality of formats takes care of that.

Thinking about modes of communication and developing alternative vocabularies is necessary because language creates obstructions in multi- and interdisciplinary conversations. In recent years I have been a participant in many discussions at conferences and in educational institutions. It is striking how often these talks address the different meanings of words and concepts in various fields. The aim to write a text devoid of technical terminology or obscure idiom may help to clarify exchanges, although of course this cannot be solved by language alone. A transdisciplinary dialogue cannot really take place in a singular language.

Lastly, practice is involved in this research project to feed the insights of the research back into the conversations in the studio. I do not work on my own. At the moment I collaborate with six people. Some years ago I started to expand my studio for strategic reasons. I did not want to be typecast as a designer and found the prospect of working on only one type of project limiting. Having a bigger studio meant being able to both work on a bigger variety of projects and to take on more extensive and more complex projects. As it turned out, working with a group of people from various backgrounds made dialogue, a fundamental factor of any design process, a natural part of the activities of the studio and enriched my praxis. It was therefore a conscious decision to involve my studio in this investigation precisely for this reason: to make the concerns of the research part of the conversation and of shared thinking of the studio.

Theorizing the Shifting Roles of Producers and Users

For the theoretical part of my research I looked for models to describe the transformation of the fields of graphic design and cartography. As explained above, these two fields have distinct origins but their current practices are similar. The output of both fields is produced in a comparable way. And both are experiencing a similar opening up of the field as a consequence of digital technologies. The theories I was looking for needed not only to incorporate this impact of tools and technologies, but also the changing role of user and producer.

While a lot has been written on the impact of evolving technologies on the role of the graphic designer, less has been theorized about the shifting position of the user. I found little consensus about the origins of the field of graphic design. Depending on whether the emphasis was on the design or the designer resulted in a very different genesis of the discipline. It strengthened my conviction that there was room for developing a new model of the evolution of the graphic design field. In this respect the work of American author, curator and graphic designer Andrew Blauvelt was important. Blauvelt connects the transformation of the role of the designer to changes in graphic design tools.[4] The model I propose defines distinct sets of technologies that enable different practices of production and use of graphic information.

In post-representational cartography, a term coined by British-American geographer John Pickles,[5] I found an approach that focuses on the full scope of map creation and production, including the role of the user. According to post-representational cartography, a map is never fully formed, it is constantly being produced and reproduced, every time a user engages with it.

What I take from post-representational cartography is the consideration that "making" and "using" are not consecutive processes but parallel tracks. Furthermore, post-representational cartography provides ways to critically think about a variety of practices engaged with cartography beyond those of professional specialists. I appreciate the shift of focus from the end product to "the work that maps do, how they act to shape our understanding of the world, and how they code that world."[6]

Exhausting and Scraping as Discursive and Artistic Methods

This book is an investigation in the domain of artistic research, a field that reflects critically and theoretically on artistic and societal issues in practice and in writing. In this introduction I have discussed many of the individual components of the above definition, perhaps without clearly presenting a comprehensive idea of how to combine them into an overall approach for this book. Before I can boil it down to such a strategy, I have to introduce one more ingredient: the overwhelming attractiveness of the map.

3   Venturi, Scott Brown and Izenour, *Learning from Las Vegas: The Forgotten Symbolism of Architectural Form*, 3.
4   Blauvelt, "Tool (Or, Post-production for the Graphic Designer)."
5   Pickles, *A History of Spaces: Cartographic Reason, Mapping and the Geo-Coded World*, 160.
6   Ibid., 12.

JG   Joost Grootens (NL) studied architectural design at Gerrit Rietveld Academy Amsterdam, and artistic research at PhDArts, The Hague and Leiden, established Studio Joost Grootens (SJG) in 1995.

LU   Linda Ursem (NL), office manager, works at SJG since 03/2012.

DJ   Dimitri Jeannottat (CH), senior designer, studied graphic design at Hochschule der Künste Bern, works at SJG since 03/2013.

SK   Silke Koeck (AT), senior designer, studied graphic design at ECAL, Ecole cantonale d'art de Lausanne, worked at SJG 11/2013–02/2019.

JS   Julie da Silva (FR), senior designer, studied graphic design at ÉSAAB, École Supérieure d'Arts Appliqués de Bourgogne, Nevers—Université Rennes II, Rennes, works at SJG since 08/2015.

CS   Carina Schwake (DE), designer, studied communication design at University of Fine Arts, Saar, worked at SJG 08/2017–06/2019.

MA   Megan Adé (CH), designer, studied visual communication at Academy of Art and Design Basel, works at SJG since 09/2018.

SR   Simon Ruaut (FR), intern, studied graphic design at École de recherche graphique, Brussels, and Royal Academy of Fine Arts, Ghent, worked at SJG 09/2016–03/2017.

SB   Salomé Bernhard (FR), intern, studied graphic design at École Nationale Supérieure d'Art de Nancy, worked at SJG 09/2016–03/2017.

YK   Yulia Kondratyeva (RU), intern, studied graphic design at State University of Printing Arts, Moscow, worked at SJG 09/2017–11/2017.

RM Raphael Mathias (DE), intern, studied communication design at Hoch-
    schule für Kunst, Design und Populäre Musik Freiburg, worked at SJG
    01/2018–07/2018.
SS Stella Shi (CN), intern, studied graphic Design at Willem de Kooning Acad-
    emy, Rotterdam, worked at SJG 02/2019–06/2019.
DV Denisse Vega de Santiago (MX), intern, studied arts and culture at Leiden
    University, worked at SJG 01/2019–05/2019.
CG Clémence Guillemot (FR), intern, studied graphic design at École Supé-
    rieure d'Art et de Design, Amiens. Works at SJG since 08/2019.
LO Laura Opsomer (FR), intern, studied type design at Masters école Estienne,
    Paris, worked at SJG 01/2020–06/2020.

Ever since I started working with maps, I noticed that users are blinded by the beauty of maps, or by the underlying worlds and processes to which they refer. Historical maps, in particular, seem to possess the power to captivate viewers beyond reason. People whose intellectual judgment and capacities I hold in high regard, approach me and tell me that maps are "really their kind of thing" without providing any further explanation, contextualization or the slightest hesitation, as if the message in itself answers all questions.

As a mapmaker, I still see the aesthetic qualities of maps and their ability to transport viewers mentally to other places. Simultaneously, and more and more, I see the map as a graphic typology in which production, visuality and information coincide to the maximum. Many things concur in the map: data is presented in a multilayered way and in great density. The data is transformed into an image that speaks to viewers on more levels than just the intellectual. And this is done in a way that explores the limits of production technologies in the sense of maximum detail and a multitude of colors. As I see it, the map is therefore a format that is especially appealing to graphic designers because so many aspects of the field coincide in the production and use of maps. However, if I want viewers to look at maps the way I see them, I cannot do that by showing them a map. They need to un-see the map, look beyond the beautiful image or the fascinating worlds it refers to, before they can view it as a graphic typology to investigate production, visuality and information. Therefore, maps are not included in this book about mapmaking, as they would distract the reader.

The insights and considerations I presented above have brought me to a combined method of unfiltered multitude and concentrated abstraction, which I call "exhausting and scraping." These double but opposite approaches fit a project in artistic research in which the discursive and the artistic are intertwined flows of investigation that have been developed independently but in parallel.

I call the method used in the discursive part of the research "exhausting," in reference to *An Attempt at Exhausting a Place in Paris* (*Tentative d'épuisement d'un lieu parisien)*, a book by French novelist Georges Perec (1936–1982).[7] In the book's forty-seven pages, Perec describes the things that happen at Place Saint-Sulpice in Paris during two days in October 1974. Rather than focusing on notable things, he outlines things that usually go unnoticed. Perec does not describe the police station, the three cafés, the movie theatre, the church, the travel agency, the bus stop or the fountain on the square, but instead lists the rest: "That which has no importance: what happens when nothing happens."[8] Perec uses the word *épuiser* to describe his method, which can be translated as "to exhaust": to use up, to wear out and to say all that can be said about a subject.

In the discursive part of this research I intended to gain new insights through analysis and detailed description of the output, process, production and context of the three mapmaking practices of nonspecialists. To learn from these contemporary mapmakers it is necessary to withhold judgment, to not bring a preformulated idea to the table, but to compare their practices with a variety of theories, as well as with experiences from my own practice.

Unlike Perec, my method of exhausting is not based on observation alone. In each case study a different set of theories is used to investigate that particular practice. In Chapter 3, different theoretical approaches to cartography are used to investigate Google Maps. In Chapter 4, a variety of perspectives is employed to study the Strava Global Heatmap: economic, technical, social and cultural viewpoints. In Chapter 5, I compare the practice of amateur mapmaker Thomas van Linge with those of specialist mapmakers in order to gain a better understanding of Van Linge's Syria maps, of how they are produced and what they represent.

I call the method used in the artistic part of the research "scraping." *Scraping* is a term used both for the extraction of data from a website, and for the technique German visual artist Gerhard Richter applies in his abstract paintings in using a squeegee to scrape paint across the canvases.[9] It could be argued that in both cases scraping is used to turn information into data. In digital scraping the content of web pages is extracted and stripped of its design so it can be used as data. In the case of Richter, the original image disappears as a result of scraping and what remains is a new (uncontrolled) organization of matter, of paint. Richter uses a similar technique as in his photo-paintings. In the final step of the process he blurs the image by means of a soft brush or squeegee. Richter has described this as blurring out "the excess of unimportant information."[10]

For the artistic part of the research I have developed a series of visualizations—timelines, diagrams, illustrations, legends—that offer alternative and an additional representation of the research. Their content is similar to that of the discursive part. The visualizations show the maps without showing them as map, but rather as graphically produced representations of information that is embedded in a technological, political and cultural context. These visualizations explore the space on the page between text and map, between text block and image. One type of visualization that is used is the timeline in which the horizontal position of words refers to a moment in time. The mathematical use of white space on the page as an added dimension—time—creates a more spatial constellation of words than that of a linearly constructed text block. Other visualizations focus on the use of colors in maps. They consist of large page-filling color blocks that are between a legend and a map, between a list of definitions and the application of those definitions to turn data into information. The purpose of diagrammatization and abstraction in the visualizations is to focus the viewer's gaze and avoid distraction.

What appeals to me in exhausting and scraping is that both terms refer to a specific approach to creating information—saying all that can be said, and distracting data from a source—while both also refer to highly physical processes—wearing out, and removing matter from a surface by applying a hard object. This specific duality between seemingly opposing qualities is one of the many I encountered in this research.

The Book as Discursive and Artistic Format

For this dissertation I have chosen the format of the book to present the artistic and academic output of my investigation. This is an artistic research project, a type of investigation that usually involves two outputs, a dissertation and an artistic presentation. This book combines both. This decision has been informed both by my practice and the research subject.

The book is my artistic weapon of choice. It has been the main output of my design practice over the past twenty years and it is therefore the format I am most familiar with. But there are other reasons. Whereas a few years ago I felt I had to

7   Georges Perec, *An Attempt at Exhausting a Place in Paris*.
8   Ibid., 3.
9   Richter's method can be observed extensively in German director Corrina Belz's 2011 documentary *Gerhard Richter Painting*.
10 Richter and Obrist, *The Daily Practice of Painting: Writings and Interviews, 1962–1993*, 37.

Projects

☐ Laus

◇ Museum Index, Van Abbemuseum, Eindhoven

⬡ Van Dale Groot woordenboek van de Nederlands

○ De ruimtelijke metamorfose van Nederland
⬤ Young–Old
○ Beyond the New
⬤ Informal Market Worlds Atlas
○ Informal Market Worlds Reader
○ Honsinzi House
○ Reproducing Scholten & Baijings

⬤ Brick: An Exacting Material
⬤ Jan Schoonhoven
⬤ Van Dale Groot woordenboek van de Nederlands
○ MVRDV Buildings (2nd edition)
○ A Man, A Village, A Museum. Li Mu: Qiuzhuang Project
○ 30:30 Landscape Architecture

Research

PhDArts

Teaching

DAE
KADK
ISIA

Lectures / Juries / Workshops

✕ TUM, Munich
✕ Willem de Kooning Academy,
✕ Integrated, Antwerp
✕ Van Piere, Eindhoven
✕ WEI SRAUM, Innsbruck
✕ ETH, Zurich
✕ AGI Open, Biel/Bienne
✕ Jan van E
+ Fedrigoni Top Award, London
+ ETH, Zurich
+ BNO Piet Zwart Prize, Amsterdam

⫽⫽ HEAD, Geneva

Studio

AD

EM
JS

LN   MB   MS
SK
DJ
HS
LU
JG

2015

2016

This timeline shows the activities studio SJG developed parallel to this research.

◇ Dream Out Loud, Stedelijk Museum Amsterdam, exhibition graphics

◇ Control Syntax Rio, Het Nieuwe Instituut, exhibition graphics

◇ #tvclerici, Milan, exhibition

◇ The 1980s, Van Abbemuseum, Eindhoven, infographics

Het ABC van de Collectie, Stedelijk Museum Schiedam, infographics

○ L'Internationale, Ma

○ DAE, #tvclerici

○ DAE, Master departments

aal

○ Atlas of Design Tool

○ Towards a Process of Makin

○ After Baldus: Travels in a Wounded Landscape

○ The Story of Art

○ Structures in Building Cultur

○ Elemental Living

○ Building Tectonic Structures

aal

○ Findings on Light

○ The Public Interior as Idea and Project

○ Notes on Ghosts, Disputes a

otterdam

× National Design Centre, Singapore

× ESAD, Porto

× FHNW, Basel

× Future Cities Laboratory, Singapore

× ECAL, Lausanne

× Exemplaires, Strasbourg

× Meermanno, The Hague

× Paju Book City, Paju

× Academy of Architecture, Amsterdam

× Academie, Maastricht

× Dutch Design Week, Eindhoven

+ BNC

⫽ ECAL, Lausanne

⫽ ZHdK, Zurich

╱ Paju Book City, Paju

CJ

SR

SE

SB

LT

2017

○ Book
◇ Exhibition
□ Poster
○ Website

○ Award-winning

× Lecture
+ Jury
╱ Workshop

AD Alexandre Debelloir
CC Charlotte Carletto
CG Clemence Guillemot
CJ Chen Jhen
CS Carina Schwake
DJ Dimitri Jeannottat
DV Denisse Vega de Santiago
EM Elena Meneghini
HS Hanae Shimizu
JG Joost Grootens

JS Julie da Silva
LN Luca Napoli
LO Laura Opsomer
LT Lorenzo Toso
LU Linda Ursem
MA Megan Adé
MB Mateo Broillet
MS Matthieu Salvaggio
RM Raphael Mathias
SB Salomé Bernhard

SE Seo-Kying Kim
SK Silke Koeck
SP Shirin Pfisterer
SR Simon Ruaut
SS Stella Shi
YK Yulia Kondratyeva

defend my choice to make books, as it was perceived by many to be an outdated format, today there seems to be change of sentiment. In a time of disinformation, misinformation and fake news, websites and other digital formats are regarded as democratic and dynamic, but also as vulnerable and volatile. Moreover, in a context where the data describing our lives is abundant and in flux, the book as an edited, well-considered, static format allows its users to step outside of the flow of information and reflect on it from the side-lines. To make a book in our day, it seems, has become a political choice.

Another reason for making a book as the artistic outcome and as embodiment of the research process is motivated by the subject of the research itself. The book format optimally reproduces critical reflective texts and visual research. In addition, the maps I investigate can only be found online and are ephemeral by nature. They are constantly being updated or are in danger of getting lost because changing technological contexts make them less relevant and therefore their creators lose interest in keeping them. Reproducing the maps in this book is a way to record them, although I am aware that they cannot be separated from the moment of use, the specific application and the device on which they are displayed. Furthermore, I have selected these maps for their design value, for how they look and function. Technology companies and amateurs may not be inclined to document their maps for their design traits, because they do not recognize their visual qualities, effective functionality or ingenious production strategy. By contrast, the book is a stable format that enables the capturing of a practice. Moreover, the static book emphasizes one of the aims of my research, which is to pause and carefully study the maps.

My design practice has taught me that it is easier to display a sign, object or space when the medium that displays it differs significantly from the original. It is much harder to show a book in a book than to document a building in printed matter. Apparently, if the reproduction is more remote from the original it is easier to conceptualize the representation process. This insight is another reason to choose the book as format for documenting screen-based mapmaking practices.

The visualizations in this book might look like what American visual theorist and culture critic Johanna Drucker has called Trojan horses from the empirical sciences.[11] The specific look of the visualizations was partly the result of the method of scraping that I described above.

The empirical character of some of the visualizations is informed by the "mixed methods" approach that combines different kinds of data in a single display. The collecting of qualitative and quantitative research data, and their integration and organization in a specific visual presentation, has been explored in a variety of fields under the name "mixed methods."[12] Mixed methods, and specifically the map as a display that combines various kinds of data, were explored at the "Mappings as Joint Spatial Display" conference in Berlin in late 2018, where I gained insights into this approach.[13] The conference's aim was to start an interdisciplinary methodological conversation at the intersection between sociology of space, architecture, urban studies and geography, about the instrument of mapping for research of spaces. Two specific and disconnected methods of data integration were mixed: mapping from the spatial sciences, and the joint displays from sociology in which qualitative and quantitative information are visually presented in a single display.

What appeals to me in the mixed methods approach is the integration of different kinds of data. The method integrates the complex data of a few cases typical of qualitative research (such as the in-depth analysis of the practices in Chapters 3, 4 and 5) with the low-complexity data of many cases that are typical of quantitative research (such as the comparison of a large variety of graphic design tools, practices and histories in Chapter 2). The combination of these different approaches allows me to zoom in and out, from the long-term trends of the larger field to the peculiarities and specifics of an individual practice. Furthermore, the mixed methods approach uses both textual and visual research formats that allow me to explore visuality, one of the key aspects of the research subject. By combining text and visualizations, I can also address another ambition I have: to develop alternative and additional languages and modes of communication to reevaluate graphic design.

The discursive part—text—and the artistic part—visualization—are not shown together on the same page. Instead, they are presented on alternating double pages. Over the past twenty years I have designed over 200 books, the majority of which contain both texts and images. As a book designer, I am familiar with many different ways of organizing visual material and typography on the page, emphasizing the one, or favoring the other kind of content. In my experience, a complete balance between text and image material can never be achieved, as each reader has a different disposition towards one kind of content over the other. Besides, I regard the discursive and artistic part as two separate argumentations, the one textual and the other visual, based on the same research. I want to emphasize that these two lines of reasoning are equal by giving them the same amount of space, literally. In terms of production, the book is built up of folded and cut sheets that are printed on one side with the texts, and with visualizations on the other. The paper has a certain translucency that allows the visual information on the one side of the sheet to shine through on the other. So while looking at texts, images can be seen as ghost-like presences. The same goes for the opposite. On the pages with the timelines, diagrams and illustrations, the lines of text shimmer through.

In graphic design, the degree to which information printed on the back side of paper shines through on the front side is, paradoxically, called opacity. Thicker paper that is less transparent than thinner stock has a higher percentage of opacity. In my book designs I like to work with paper that is not completely opaque, but slightly shows the texts and images on the back of the page. If the paper is too transparent the information on either side of the page becomes illegible. But if it only just shines through, the information on a page becomes connected to the rest of the book. It de-emphasizes a particular page and instead highlights the book as a coherent information system. This semitransparent quality makes a page into a spatial object rather than a flat image. I make use of translucency as a visual device in this book to connect the discursive and the artistic, the textual and the visual. It is also my aim to use translucence as a conceptual device to examine a variety of practices, to put them on top of each other, hold them to the light, and compare their silhouettes. To compare graphic design with cartography and specialist practices with those of nonspecialists. My own practice, which has aspects of all of these, will be used to look at the processes and products of others. All with the aim of learning from diverse practices to transform my own.

11 Drucker, *Graphesis: Visual Forms of Knowledge Production*, 125.
12 Creswell and Plano Clark, *Designing and Conducting Mixed Methods Research*.
13 Collaborative research center "Re-Figuration of Spaces," "Mappings as Joint Spatial Display," accessed 12 April 2019, https://www.sfb1265.de/veranstaltungen/conference-mapping-as-joint-spatial-display/.

Projects

raphics

◇ Mined, Eindhoven, exhibition graphics
□ Mined

ping Collections

○ AEIOU: Articles, Essays, Interviews and Out-takes by Tony Fretto
○ Living on Water   ○ Design Academy Eindhoven: M
○ Design Academy Eindhoven. Master Course
○ Ornament and Identity: Neutelings Riedijk Architecten
● The Noise Landscape: A Spatial Exploration of Airports and Cities   ○ Textbook: Kees Christiaanse
III: Steel Skeleton   ● Mined: Graduation Catalogue 2017 Design Academy Eindhoven   ○ Potato Plan Collection
○ A State Beyond the State: Shenzhen and the Transformation of Urban China
: Crafting Wood   ○ This Baby Doll Will Be a Junkie   ○ Van Dale Pocketwoordenboeken
nd Killer Bodies   ○ Boragó: Coming from the South   ○ Objec
○ Future Cities Laboratory: Indicia 01   ● Atlas of the Copenhagens   ○ Questioning D

Research

PhDArts

Teaching

DAE
KADK
ISIA

Lectures / Juries / Workshops

✕ De Pastoe Fabriek, Utrecht   ✕ HKB, Bern
✕ Jacques Bertin et le Laboratoire de graphique, EHESS, Paris
✕ AGI Open, Paris   ✕ ELISAVA, Barcelona
                                    + GRA-prijzen, Rietvel
wart Prize, Amsterdam   + CHEOPS design competition, Noordwijk
          ⊞ Most Beautiful Swiss Books, Bern   + Most Beautiful S
                    ⫻ HKB, Bern

Studio

CJ   CC
JS
SE   YK   RM   MA
SK
DJ
LT   CS
LU
JG

2018

◇ Museum Index, Van Abbemuseum, Eindhoven
◇ G18, Eindhoven, exhibition graphics
□ G18

ter Graduation 2018　　　　○ The Wandering Maker
　　　　　　　　　　○ Future Cities Laboratory: Indicia 02
　　　　　○ Living in the Desert
　　○ Crafting the Façade: Stone, Brick, Wood　　　○ Writing to Louis Andriessen
Spots in Shots　　　　　　　　○ Parliaments of the European Union　　　　　○ Atlas of Mid-Centu
Atlas of Brutalist Architecture　　　○ Van Dale Groot woordenboek voor school　　○ Architecture and Argument
G18: Graduation Catalogue 2018 Design Academy Eindhoven　　　　○ Een huis genaamd Marseill
f Nederland: Veranderd landschap 1974–2018　　　　　　○ Shared Cities Atlas
ign: The Master Programmes at Design Academy Eindhoven　　○ The Grand Projet

⸭ Visualizing Knowledge, Aalto University, Helsinki
× Graphic Hunters, Utrecht
× VI PER, Prague
× FHNW, Basel　　　　　　　　　　　　× The Ber
× Future Urban Regions, Amersfoort　　　　× FH, Aachen
× Mappings as Joint Spatial Display, HKW, Berlin　　× AGI Open, Rotterdam
cademie, Amsterdam

+ BNO Piet Zwart Prize, Amsterdam
s Books, Zurich　　　⧻ Most Beautiful Swiss Books, Bern　　⧻ Most Beautiful Swiss Books, Zurich
／ Mappings as Joint Spatial Display, HKW, Berlin

SS
P　　DV

LO

CG

2019　　　　　　　　　　　　　　　　　　　　　　　　2020

2   Positioning

In this chapter I will try to gain an understanding of the transformations that graphic design has gone through in the digital age. I propose a "model of technological thresholds," a sorted timeline of design tools, to better understand the ever-evolving transformation of the field. Building on experiences from my design practice and applying theories on graphic design and cartography, I will draw parallels between shifts happening in both the design of graphic information and in map-making that I believe to be fundamental. Using ideas from post-representational cartography, I will introduce the concept of the blind map to describe the blurring of the producer–user divide in the production of visual information. I will conclude the chapter by outlining methods and criteria for further research.

This chapter's title refers both to technologies that enable the identification of geographical locations, such as in the Global Positioning System, and to the general question of this research to determine my own position in various fields.

Graphic Design as Activity, Output and Field

The term *design* is used both as a verb to describe an activity and as a noun to define its output. Accompanied by an adjective it can also outline a field: industrial design, for instance.

Designers tend to give very open definitions of the output of their activities. "Design" is usually understood as the output, any output according to some, of the designer.[1] British graphic designer and author Richard Hollis states that "not only is the activity called design, but design is the outcome and the expression of what a designer does."[2] At the Design Academy Eindhoven, the school where I teach, "a" design is regarded as any work created by a designer: a strategy, a system, a book, an installation, a film, clothing, a fabric, a machine and more.[3]

As a designer, I find these definitions of design both too limited and too open. To me, design is not a field restricted to those with a specific training or specialized professional practice, in other words, design is not the exclusive domain of designers. The democratization of design tools in the digital age has opened up the field to virtually anyone and resulted in designers no longer being the sole creators, editors and producers of design. At the same time, I think there are certain limitations to what design can do.

Design is not one of the so-called autonomous arts. Traditionally, the impetus for a design comes from an outside source. And although nowadays we may consider design more broadly to also initiate a design process and speculating without a direct question or commission, it remains the case that design responds to changes in production technologies, economic conditions and insights about concrete use or application. Design will therefore never be completely autonomous.

In this research project I will concentrate on graphic design, a field that focuses on the editing and production of visual information that is multiplied by reproduction methods. Information here is understood as transformed, edited or organized data. Data are things known or assumed as fact that do not have a built-in visual form, like a text, list or database. Once given form—shape, color, typeface, size, composition—this data can become information. I regard knowledge as the understanding and interpretation of information through experience or education.

1   In this book I will use the pronoun "she" to designate graphic designers to express that graphic design has a diversity of practitioners. This is something I see in my working environment: the majority of design students I teach are female, and so are most of the collaborators in my studio. More importantly, I want to recognize the gender imbalance in the descriptions and historiographies of graphic design. This is an insight that I obtained from reading Catherine De Smet's article "*Pussy Galore* and Buddha of the Future: Women, Graphic Design, etc."

2   Hollis, "Have You Ever Really Looked at This Poster?," 73.

3   Knowledge Circle Design Academy Eindhoven, "Lexicon of Design Research," definition of "design," accessed 5 October 2018, http://www.lexiconofdesignresearch.com/lexicon/texts/design.

1

2

Quantitative research has been carried out into recent historical descriptions of graphic design. To this end, descriptions have been selected that describe a general global history, and not a particular area or period. For practical reasons, English-language books were chosen. This resulted in a selection of five books on the history of graphic design that not only differ in terms of the definition or approach of the field, but also in terms of the format and size of the book as shown by these scans of the book's bottoms. They are: Drucker & McVarish, *Graphic Design History: A Critical Guide*, 2nd edition, 2013 (1); Eskilson, *Graphic Design: A History*, 2nd edition, 2012 (2); Hollis, *Graphic Design: A Concise History*, 2001 (3); Jubert, *Typography and Graphic Design: From Antiquity to the Present*, 2006 (4); Meggs & Purvis, *Meggs' History of Graphic Design*, 5th edition, 2012 (5).

3

4

5

Using typography, illustration and layout, graphic designers create, edit and combine symbols, images and text to visually represent ideas and messages.

Recent attempts at new names for and definitions of activities of designers include relational design and critical design.[4] They appear to be based on the assumption that terms like industrial design and graphic design are too limited and merely indicate a simple service-oriented industry. British/Australian graphic designer, educator and author James Goggin argues that designers, design critics and historians should broaden their perception of what exactly graphic design encompasses, and that they should be aware of the unique position it occupies between reading, writing, editing and distribution.[5] As a discipline it is nuanced and expansive enough in its everyday activities and processes to make renaming unnecessary, according to Goggin.

Goggin's argument is valid, in my opinion. The term *graphic design* is appropriate for the activities it covers. The term *graphic design* is appealing because it refers to the industrial production of visual information. I would find it problematic if the term *design* were to suggest a certain exclusivity in terms of training or professional status. If design is what designers do, than this does not automatically imply it can only be done by those with a specific education or specialized practice.

The origins of graphic design date back to the nineteenth century. A proto-version of the profession emerged as part of the activities of printing and publishing houses. With the invention of color lithography, illustration and text drawn by hand could be joined on the same printing surface. This innovation gave a boost to the artistic and technological development of color lithograph posters, printed separately for each color and in multiple editions, which were simultaneously distributed throughout the urban environment.[6] The producers of these posters identified themselves as commercial artists or graphic artists, who were responsible for each element in a design intended for reproduction by machine. They were practicing what was later recognized as graphic design.[7] A further evolution of the field came with the development of black-and-white photography. Flashbulbs, faster film emulsions and lenses made photography the dominant pictorial technique, replacing drawn and painted illustrations.[8]

In Europe, commercial artists were joined by avant-garde artists who regarded graphic design as a means of extending art into modern life. These "artist-designers" or "painter-graphic designers," as they called themselves, exploited photography and exposed new meanings by juxtaposing images. At the same time they subverted it, destroying and reassembling its images through montage, which was to become a new expressive and critical device. Informed by the ideas and works of futurism and its disdain for tradition, the artists of the Dada movement—anti-establishment, anti-military, anti-art—continued the revolution in the use of words and images, as they mixed all kinds of letterforms and printers' ornaments in typographic compositions.[9] The developments from expressionism to functionalism, and from handicraft to design for machine production, mark the end of the process of specialization and emancipation of graphic design. In my view this is the beginning of graphic design: when it describes both a specialized activity, distinct from production and a field of autonomous professionals constituting a network.[10]

From the early days of graphic design until the digital age, the position of the graphic designer in the information chain, and the successive steps in the process of creation, production, dissemination as well as the use of graphic information have basically remained the same. The graphic designer was a specialist who edited visual information and prepared it for reproduction. The advent of the computer and design software fundamentally changed the position of the designers and users of graphical information.

Hollis has described the history of graphic design as the history of the designer taking control of the craftsman's process.[11] That trend culminated in the introduction of the Apple Macintosh Computer (1984) and desktop publishing. The industrial production of graphic information had established a series of specialist activities like layout, typesetting, lettering and color separation. The desktop publishing revolution brought these activities into the practice of graphic design, discarding many production jobs in the process. As it did with other professions in the information chain, the computer as a tool changed the practice of graphic designers, adding production tasks to their conceptual responsibilities. On the one hand this gave them more control than ever before, but at the same time it increased the workload considerably, as they now had to do more and different kinds of work. American graphic designer, curator and author Ellen Lupton, referencing Walter Benjamin's 1934 text "The Author as Producer," has described this shift in practices as the "designer as producer."[12]

Walter Benjamin claimed that the borders between writing and reading, authoring and editing were dissolving in new forms of communication like film, radio and the illustrated press. To bridge this divide is a revolutionary act: "The barriers imposed by specialization must be breached jointly by the productive that they were set up to divide."[13] Benjamin condemned the model of the writer as an expert who only creates texts and is not aware of the physical life of the work. Instead, Benjamin proposed the model of the producer who questions where a work will be read, by whom, what other information will surround it, and how it will be manufactured. All with the goal of turning users "into producers—that is, readers or spectators into collaborators."[14]

Lupton's 1998 text "The Designer as Producer" was a response to the idea of the "designer as author" that emerged in discussions about graphic design in the 1990s. The recognition that information is not neutral and that its presentation shapes how the user perceives content made designers into more than the functional service providers modernism thought them to be. In the context of the rise of star architects like Rem Koolhaas and star designers like Philippe Starck, the graphic designer's wish for recognition might be understandable. However, the concept of the author is a problematic one, following the critical writings about the author as authority figure by Barthes,[15] Foucault[16] and others. According to British author, lecturer and curator Rick Poynor "the very notion of an 'author' as a validating source of authority for a cultural work is outdated, backward-looking and reactionary."[17] American graphic designer Michael Rock concludes, in the 1996 text "The Designer as Author," that, except for a very few examples, the authorship model is not adequate as a way of thinking about design. He suggests three

4   The term *critical design* was first used by British designer Anthony Dunne of design studio Dunne & Raby in his book *Hertzian Tales: Electronic Products, Aesthetic Experience, and Critical Design* to describe a kind of design that is initiating and speculating without a commission but rather creates scenarios, raising "what if" questions about the future and critiquing of contemporary societal issues.

5   Goggin, "Practice from Everyday Life: Defining Graphic Design's Expansive Scope by its Quotidian Activities," 55.

6   Boekraad, "Graphic Design as Visual Rhetoric: Principles for Design Education," 7.

7   Hollis, *Graphic Design: A Concise History*, 16.

8   Ibid., 38.

9   Ibid., 52.

10 This reading of the origins of graphic design is not uncontested. The debate of the history of graphic design can be found in the collected articles in De Bondt and De Smet, *Graphic Design: History in the Writing (1983–2011)*. A comparison of five books on the history of graphic design makes clear that there is no consensus about the start of the field (Drucker and McVarish, *Graphic Design History: A Critical Guide*; Eskilson, *Graphic Design: A History*; Hollis, *Graphic Design: A Concise History*; Jubert, *Typography and Graphic Design: From Antiquity to the Present*; Meggs and Purvis, *Meggs' History of Graphic Design*). Depending on whether the emphasis is on graphic communication, on the graphic object or on the graphic designer the start of the field varies from prehistory until the early twentieth century.

11 Hollis, "Have You Ever Really Looked at This Poster?," 73.

12 Lupton, "The Designer as Producer," 13.

13 Benjamin, "The Author as Producer," 87.

14 Ibid., 89.

15 Barthes, "The Death of the Author," 142.

16 Foucault, "What Is an Author?," 13.

17 Poynor, *No More Rules: Graphic Design and Postmodernism*, 118.

1

From Prehistory to Early Writing

3

Classical Literacy

15

Medieval Letterforms and Book Formats

30

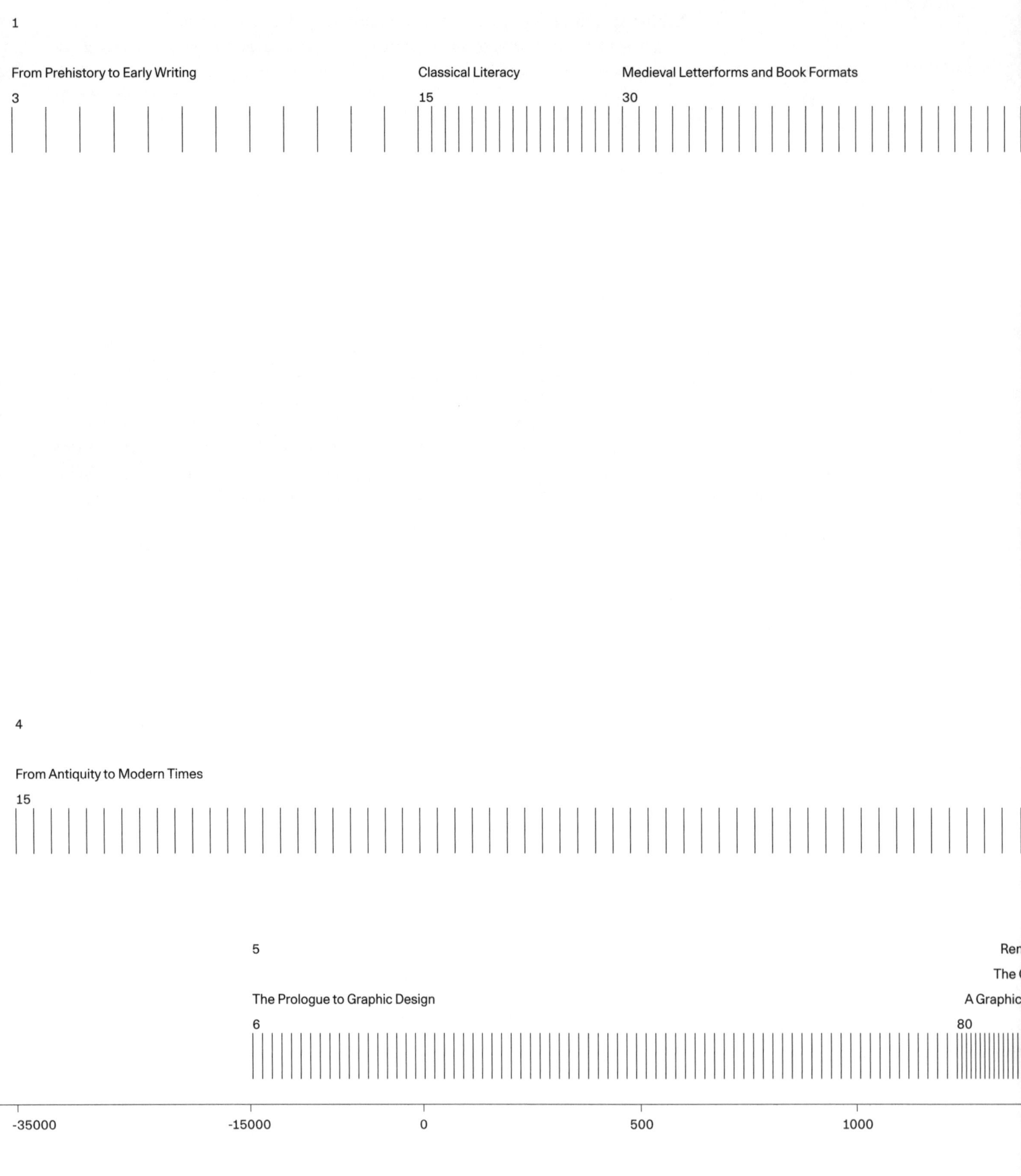

4

From Antiquity to Modern Times

15

5

The Prologue to Graphic Design

6

Ren

The C

A Graphic

80

-35000   -15000   0   500   1000

This timeline compares the five selected graphic design histories. The pages of the books are plotted on a timeline in the form of dashes. Each line represents a page. More extensively discussed periods have more dashes. The pages of chapters are evenly distributed on the timeline for the period that is addressed in the chapter. The comparison shows that there is no consensus among the different historiographies about the beginning of the field. Also, each description emphasizes different periods.

Formations of the Modern Movement

Mass Mediation
Postmodernism in Design

The Graphic Effects of Industrial Production
Pop and Protest

Modern Typography and the Creation of the Public Sphere
Public Interest Campaigns and Information Design
Digital Design

enaissance Design: Standardization and Modularization in Print
The Culture of Consumption
Graphic Design and Globalization

80 105 125 150 200 220 245 285 310 345

Modern Art, Modern Graphic Design

Revolutions in Design

Sachplakat, the First World War, and Dada

The Bauhaus and the New Typography

Art Nouveau: A New Style for a New Culture

The Nineteenth Century: An Expanding Field
The Triumph of the International Style
Contemporary Graphic Design

e Origins of Type and Typography
American Modern and the Second World War
Postmodernism: The Return of Expression

25 60 105 140 210 240 285 320 370 430

War and Propaganda

The Avant-Garde and the Origins of Modernism

3
National Tendencies in Europe
New Waves: Electronic Technology

The United States
Psychedelia, Protest and New Techniques

From Graphic Art to Design
Variants of Modernism in Europe
From 1990 to the New Millennium

10 30 110 180 185 215 225

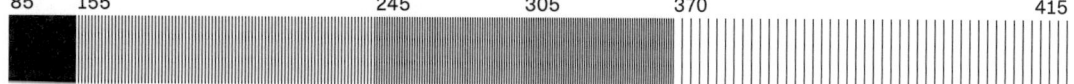

The Final Decades of the Twentieth Century

The First Third of the Twentieth Century
Postwar Economic Boom

The Long Nineteenth Century
From the Rise of Totalitarianism to World War II

85 155 245 305 370 415

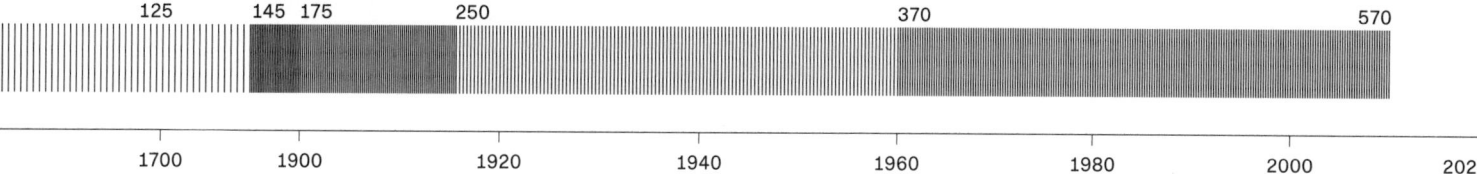

The Modernist Era

The Arts and Crafts Movement and Its Heritage

ce Graphic Design
The Bridge to the Twentieth Century

n Illustrated Book
Graphic Design and the Industrial Revolution

ssance
An Epoch of Typographic Genius
The Age of Information

125 145 175 250 370 570

1700 1900 1920 1940 1960 1980 2000 2020

1 Drucker & McVarish, *Graphic Design History: A Critical Guide*, 2nd edition, 2013
2 Eskilson, *Graphic Design: A History*, 2nd edition, 2012
3 Hollis, *Graphic Design: A Concise History*, 2001
4 Jubert, *Typography and Graphic Design: From Antiquity to the Present*, 2006
5 Meggs & Purvis, *Meggs' History of Graphic Design*, 5th edition, 2012

alternative models to describe the processes usually involved in design activity: the designer as translator, the designer as performer, and the designer as director.[18] It is revealing that in doing so, analogies with other types of expertise are still needed to describe the specific activities of the graphic designer. Some ten years later, Rock revisits his ideas in the text "Fuck Content."[19] He argues that the idea of the designer as author grew out of a valuing of origination over manipulation of content. This stems from the false dichotomy between form and content and the wrong idea that form without content is just an empty shell. Rock states that design is "a kind of text itself, as complex and referential as any traditional form of content." Design is not "what it is about, it's how it is about," its content is "perpetually, Design itself."[20]

Lupton's reference to Benjamin broadens the scope beyond creation and production to include use. In "The Author as Producer," Benjamin's aim was to turn "readers or spectators into collaborators."[21] Lupton states that digital technologies have created opportunities for designers to intellectually and economically take control of the means of production and to share this control with users, "empowering them to become producers as well as consumers of meaning."[22]

American graphic designer, author and curator Andrew Blauvelt has argued that open access to computers and design software exposed the mysterious and invisible processes in creating graphic design.[23] It demystified the field, raised awareness about design and generated a broader interest that would produce more designers. This successively resulted in growing competition, lower salaries, an overwhelming amount of amateur work and an erosion of craft.[24] To Blauvelt, the computer recast the practice of graphic design as a set of digital tools. With the entry requirements for a practice consisting of a computer and software, the professional-amateur division is no longer applicable. To become a professional graphic designer means becoming good at using the tools and making a living of it.[25]

I will look closer at two texts and a lecture by Blauvelt in which he describes the transformation of graphic design in the past decades, proposes a model to describe this shift and suggests a way for graphic design to "save itself."[26]

In his 2003 text "Towards Critical Autonomy, or Can Graphic Design Save Itself?,"[27] Blauvelt describes the state of pluralism of the graphic design field in the early 2000s.[28] Following the desktop publishing and personal computing revolution of the 1980s, graphic design lacked coherence and had become so dispersed that it resembled a "vast formless body."[29] The field had expanded beyond its roots in print and had also diversified following a period of intense formal experimentation in the 1990s. These experiments questioned the predominant assumptions of the time, mainly those of modernism. Initially rooted in critical reflection, soon the motivation of the trials and exercises seemed to be aimed at promoting individual expression as an end in itself. A proliferation of design styles followed. These may have looked experimental—disregarding functionality, irrationally organized, deliberately chaotic and illegible—but in essence were not, because the experimentation, according to Blauvelt, was chiefly aimed at self-promotion and lacked a critical foundation. Blauvelt suggests that design may be able to save itself from this "malaise" of disciplinary formlessness by reclaiming a sense of critical autonomy. Autonomy not in the sense of withdrawal from the social, but as a discipline that is capable of generating substance out of its own means and processes without commissions, functions, specific materials or production methods.[30]

I wrote earlier that I do not consider design to be one of the autonomous arts. However, I agree with Blauvelt's statement about critical autonomy. Crucial to me is his notion of being able to generate significance out of the means and processes of graphic design itself. A designer can be autonomous in the sense that she can act independently of commissions, functions or production technologies, but a designer cannot act independently from design itself. To me, critical autonomy in design ceases to work when it is completely detached from a discipline and disciplinarity.

The critical autonomous practices contemplated in the 2003 text were the subject of the 2011 exhibition *Graphic Design: Now in Production* that Blauvelt co-curated with the aforementioned Ellen Lupton. First on display in the Walker Art Center in Minneapolis, USA, the show provided an overview of projects by graphic design practices from around the globe since 2000.

Blauvelt contributed the text "Tool (Or, Post-production for the Graphic Designer)" to the accompanying catalogue.[31] In this text Blauvelt introduces a model of the transformation of graphic design consisting of three realms: preproduction —production—postproduction.

Preproduction describes the activities of the graphic designer prior to the introduction of the computer and of design software. To design was to make a plan and prepare instructions for others to manufacture the graphic product.

In the realm of production, with the computer and with layout and design software to create, produce and distribute visual information, the designer assumed increased responsibilities. The designer-as-producer had more control over the production process than ever before, but also more work to do. The practice of graphic design was reshaped into a set of digital tools that were accessible for everyone with a computer.

In the realm of postproduction, graphic designers are orchestrators of tools, systems and/or formats. In these new practices, enabled by Web 2.0 technologies, the distinction between designer and user and between production and consumption is blurred. Labor is dispersed and creation is interdependent of co-creating users.[32] Whereas design in the sphere of production still carried overtones of authorship, originality and singularity, in the realm of postproduction, design represents co-authorship, reference and collectivity.[33]

In his 2013 lecture at *counter/point: The 2013 D-Crit conference,* titled "Graphic Design: Discipline, Medium, Practice, Tool, or Other?," Blauvelt looks back at both of the aforementioned texts as well as at the exhibition *Graphic Design: Now in Production*.[34] Summarizing, Blauvelt states that similar to how the tools and practices of graphic design have become appropriated and dispersed, the graphic designer herself has appropriated the roles of others and annexed various systems of production. The focus of the 2011 exhibition *Graphic Design: Now in Production* was not the expansion of the various formats that graphic design might take, but the appropriation and blurring of the boundaries between the different practices the graphic designer engages with. Blauvelt describes these new practices as the designer as author, as editor, as publisher, as producer, as entrepreneur, as

18 Rock, "The Designer as Author."
19 Rock, "Fuck Content."
20 Ibid., 15.
21 Walter Benjamin, "The Author as Producer," 89.
22 Lupton, "The Designer as Producer," 13.
23 Blauvelt, "Graphic Design: Discipline, Medium, Practice, Tool, or Other?"
24 Blauvelt, "Tool (Or, Post-production for the Graphic Designer)," 23.
25 Blauvelt, "Graphic Design: Discipline, Medium, Practice, Tool, or Other?"
26 Blauvelt, "Towards Critical Autonomy, or Can Graphic Design Save Itself?," 8.
27 Originally published in *Emigre*, no. 64 (2003). Initially a quarterly, *Emigre* was an influential graphic design magazine published in San Francisco, United States, from 1984 to 2005.
28 Blauvelt, "Towards Critical Autonomy, or Can Graphic Design Save Itself?"
29 Ibid., 9.
30 Ibid., 10.
31 Blauvelt, "Tool (Or, Post-production for the Graphic Designer)"
32 Ibid., 28.
33 Ibid., 26.
34 Blauvelt, "Graphic Design: Discipline, Medium, Practice, Tool, or Other?"

East

Gustav Klutsis [1,2,3,4,5]

Ryuichi Y

Alexander Rodchenko [1,2,3,4,5]

Vladimir Mayakovsky [1,3,4]

Vilmos Huszar [1,3,5]   Stenberg Brothers [1,2,3,4,5]

Kasimir Malevich [2,3,4,5]   Alexey Brodovitch [1,2,3,4,5]

László Moholy-Nagy [1,2,3,4,5]

El Lissitzky [1,2,3,4,5]

Wald

Roman Cieslewicz [3,4,5]

Ardengo Soffici [2,3,5]   Karel Teige [3,4,5]

Jan Lenica [3,

Filippo Marinetti [1,2,3,4,5]

Ladislav Sutnar [1,3,4,5]

Leonetto Cappiello [2,3,4,5]

Franco Grignani [1,3,4]

Giacomo Balla [2,4,5]   Fortunato Depero [1,2,3,4,5]

Giovanni Pintori [3,4,5]

Josef Hoffmann [2,3,4]

Alfred Roller [1,2,3,4,5]

Gustav Klimt [1,2,3,5]

Herbert Bayer [1,2,3,4,5]

Julius Klinger [2,3,4,5]   Hannah Höch [1,2,4,5]   Otto Neurath [1,3,5]

Koloman Moser [1,2,3,4,5]   Hugo Ball [2,4,5]   Paul Renner [2,3,4,5]

Hans Rudi Erdt [1,2,3,5]   Joost Schmidt [1,2,3,4,5]

Peter Behrens [1,2,3,4,5]   Lyonel Feininger [2,4,5]

nas Heine [3,4,5]   Rudolf Koch [3,4,5]   Kurt Schwitters [1,2,3,4,5]   Gerd Arntz [1,3,5]

Otl Aicher [2,3,

Eckmann [1,2,3,5]   John Heartfield [1,2,3,4,5]   Anton Stankowski [1,2,3,4,5]

Adrian Frutiger [1,2,3,

Ludwig Hohlwein [1,2,3,4,5]   Josef Albers [2,3,4]

Armin H

Lucian Bernhard [1,2,3,4,5]   Jan Tschichold [1,2,3,4,5]

Siegfrie

Ernst Keller [3,4,5]   Josef Müller-Brockmann [1,2,3,4,5]

Eduar

Niklaus Stoecklin [3,4,5]   Hans Neuburg [2,3,5]   Carlo Vivarelli [1,2,3,5]

Herbert Matter [1,2,3,4,5]

Karl Ger

Johannes Itten [2,3,4,5]   Max Huber [3,4,5]   Herbert Leupin [3,4,5]

Otto Baumberger [3,4,5]   Max Bill [1,2,3,4,5]   Richard Paul Lohse [2,3,4]   Ro

Piet Zwart [1,2,3,4,5]

Ma

Henry van de Velde [1,2,3,5]   Piet Mondrian [2,4,5]   Paul Schuitema [1,2,3,4,5]

Sphane Mallarmé [3,4,5]   Theo van Doesburg [1,2,3,4,5]

Alexandre Steinlen [2,3,4,5]   Cassandre [1,2]   Paul Colin [3,4,5]

Raymond Savignac [3,

use-Lautrec [2,3,4,5]   Guillaume Apollinaire [2,3,4,5]   Jean Carlu [1,2,3,4,5]

ardsley [1,2,3,5]

Walter Crane [1,3,5]   Edward Johnston [1,2,3,4,5]   Stanley Morison [2,3,5]

les Rennie Mackintosh [2,3,4,5]   Alfred Leete [2,3,4,5]   Eric Gill [1,2,3,4,5]   Abram Games [3,4,5]   Anthor

Macdonald [1,2,3]   Saville Lumley [2,3,5]

rris [1,2,3,4,5]   Pablo Picasso [2,4,5]   Bradbury Thompson [1,3,4,5]   Saul Bas

Will Burtin [1,3,5]   Henry Wolf [3,

Man Ray [2,4,5]   Paul Rand [1,2,3,4,5]   Gene Frederico [1,

William Golden [1,2,3,5]

James Montgomery Flagg [1,2,3,4,5]   Alvin Lustig [1,2,3,4,5]   Se

Norman Rockwell [1,2,3]   Lester Beall [1,2,3,4,5]   Cipe Pineless [1,2,5]   Her

Charles Coiner [2,3,4]

West

Leo Lionni [3,4,5]

| 1900 | 1910 | 1920 | 1930 | 1940 | 1950 |

This chronological overview lists the names of graphic designers whose work is depicted in the five selected graphic design histories. Only designers who appear in at least three of the historical descriptions are included. The superscript number after the name of a designer indicates in which books of the selection of histories their work is depicted. The year of the depicted work was leading for the placement on the timeline. An average was chosen for designers with work included from several years. The vertical placement of the names is based on the designer's location, organized from east to west, from Asia via Europe to North and South America. A remarkably small number of graphic designers in this overview is female.

ashiro [3,4,5]

Tadanori Yokoo [1,3,5]

ar Swierzy [3,4,5]

Massimo Vignelli [1,3,5]

Peter Brandt [1,2,3]
Hermann Zapf [1,4,5]
Willy Fleckhaus [1,3,4,5]   Uwe Loesch [3,4,5]   Erik Spiekermann [2,4,5]
ann [1,2,3,4,5]
dermatt [1,3,4,5]
offmann [1,2,5]
Wolfgang Weingart [1,2,3,4,5]
er [2,3,4]
Emil Ruder [2,4,5]   Ralph Schraivogel [3,4,5]
arie Tissi [3,4,5]   Gert Dumbar [2,3,5]   Studio Dumbar [1,3,4,5]
iedinger [1,2,3,4,5]   Jan van Toorn [2,3,5]
Wim Crouwel [2,3,4,5]

Robert Massin [2,3,4,5]   Grapus [2,3,4,5]

Neville Brody [1,2,3,4,5]
Alan Fletcher [2,3,4,5]   Jamie Reid [1,2,3,4]   Stefan Sagmeister [1,2,4,5]
oshaug [3,4,5]   Matthew Carter [1,2,3,4,5]   Why Not Associates [3,4,5]
Muriel Cooper [1,3,5]   David Carson [1,2,0,4,5]
3,4,5   Lou Dorfsman [3,4,5]   Rudy VanderLans [1,2,3,4,5]
eorge Lois [1,3,5]   Dan Friedman [1,2,5]   Zuzana Licko [1,2,4,5]   Jeffery Keedy [1,2,3]
Victor Moscoso [1,2,3,5]   Edward Fella [1,2,3,4,5]
Barbara Kruger [2,3,4]   Elliott Earls [1,2,4]
ur Chwast [2,3,4,5]   April Greiman [1,2,3,4,5]   Katherine McCoy [1,2,3,5]
balin [1,2,3,4,5]   Peter Max [1,4,5]   Paula Scher [1,2,3,5]   Tibor Kalman [1,2,3]
Milton Glaser [1,2,4,5]   Susan Kare [1,4,5]
Wes Wilson [1,2,4,5]   Bruce Mau [1,2,3]   John Maeda [2,4,5]

| 1970 | 1980 | 1990 | 2000 | 2010 | 2020 |

1   Drucker & McVarish, *Graphic Design History: A Critical Guide*, 2nd edition, 2013
2   Eskilson, *Graphic Design: A History*, 2nd edition, 2012
3   Hollis, *Graphic Design: A Concise History*, 2001
4   Jubert, *Typography and Graphic Design: From Antiquity to the Present*, 2006
5   Meggs & Purvis, *Meggs' History of Graphic Design*, 5th edition, 2012

programmer, as archivist, as visual journalist, as tool maker, as curator, as storyteller, as educator, as artist, as researcher and as enabler.[35] To Blauvelt these practices are models of an expanded notion of graphic design that might be a way to save the field.

Friendly Fire

In my view, Blauvelt's description of graphic design in the digital age is of the graphic designer as hit by friendly fire. Friendly fire is a military term used for an accidental attack by a force on its own army or allied troops. The specialist tools that gave the graphic designer increased control and power in the digital age subsequently threaten to make her obsolete.

Blauvelt's analysis of the transformation of the field of graphic design focuses exclusively on the graphic designer. This focus is too limited. The works included in *Graphic Design: Now in Production* are predominantly made by people who were trained to be, or working as, specialized professional graphic designers, rather than the amateurs and practitioners from other disciplines who, with no prior knowledge of the field, use the tools to create, edit, produce and distribute visual information.

Even if a description of graphic design is centered on the activities and output of the designer, it is difficult to maintain that she is a singular force in the design process. Design is a collaborative activity in which the person who gives the assignment and those who produce the work play a crucial role in shaping the output. And can design be fully understood if the role of the user is not considered? Even the tools of the designer have an inevitable impact on the process: the computer and design software have incorporated specialist design and production tasks that were previously undertaken by other production specialists and designers. It is important that a model that describes the phenomenon graphic design offers room to all of these aspects.

Whereas most analyses of the nature of graphic design seem to end in the digital age,[36] Blauvelt offers a conceptual model that goes one step further. But does his model work? The preproduction—production—postproduction model is elegant in its simplicity and symmetry, but it is almost too neat. The scheme suggests that graphic design is developed in consecutive stages, but the realms of production and postproduction actually run parallel to each other. It is even questionable if postproduction should be given such a prominent place in the model as it only covers a small, albeit growing, number of practices. In addition, the prepositions "pre" and "post" of the three-stage scheme suggest it is complete. Blauvelt's categories do not leave room for imagining additional phases before or after.

What I do find interesting about the realms of production and postproduction, however, is that they are centered on complementary, but essentially distinct, sets of digital technologies. Production originates in the digitization of tools. Different tasks and tools are combined into one supertool, the computer and design software. Postproduction, on the other hand, builds on the dispersion, sharing and exchange of information via the Internet and more specifically Web

2.0 technologies. The impact of these sets of technologies on the field of graphic design is distinctly different. Digitization impacted first and foremost the tools of the designer, the tasks she had to do, and her role in the information chain. Dispersion, on the other hand, deals with (speed and expanse of) distribution of information and with interaction and exchange with others. Although different in nature, both sets of technologies reinforce each other.

I propose an alternative model to describe the transformation of graphic design: a timeline of technological thresholds. Building on Blauvelt's analysis, my model is not a textual description but a graphic representation of a period of time on which technologies are included for creating, recording, editing, producing, distributing and accessing visual information. I distinguish three sets of technologies: mechanization, digitization and dissemination. The latter two are described above. Mechanization refers to the technologies of the industrial production of graphic information that enabled the graphic designer to emerge as specialist in the production of graphical information.

The technologies in the model are collected from two books that describe the field of graphic design. They are the aforementioned catalogue *Graphic Design: Now in Production*[37] (2011) and *Graphic Design History: A Critical Guide*[38] (2013). The latter book was selected for its "tools of the trade" lists, overviews of tools employed by designers, as well as technologies used for reproduction, even if not directly by a designer.[39] I supplemented the tools, machines and software thus acquired with technologies of more recent date, appearing after the publication of both books, and selected following additional research and discussions in my studio.

The timeline itself is limited to the period 1900–2020. The starting date is related to what I see as the beginning of graphic design in the 1910s and 1920s, when it relates to both a specialized activity, distinct from production and a field of autonomous professionals constituting a network. In the timeline, the technologies are organized by date of introduction and thus not necessarily the moment they were most intensively used. The sets of technologies form three thresholds or boundaries between different spheres.

Although a timeline is a temporal structure, I read the model spatially. To me, it creates a series of consecutive conditions for different kinds of practices and formats of production and use of graphic information, each with its own levels of accessibility in terms of economics or required specialist knowledge. Nowadays, designer's practices are fluid, they move between different technological spaces, occasionally opting to produce by using predigital or even preindustrial technologies, while at other stages of a project choosing for Internet-based distribution formats. The model also allows the situating of different types of use and users: from more passive types of use to participatory forms enabled by Web 2.0 technologies, to formats of use in which the distinction between producers and users ceases to exist.

The space between, and the angles of, the different thresholds in the model, indicate the speed and scale of the transformation of graphic design. The line formed by technologies involving mechanization has a gradual angle. The other two thresholds, those of digitization and dissemination, are much steeper. This signifies that technologies involving computers and the Internet were introduced

35   Blauvelt, "Graphic Design: Discipline, Medium, Practice, Tool, or Other?"
36   The final chapter of Meggs and Purvis, *Meggs' History of Graphic Design,* published one year after Lupton and Blauvelt's *Graphic Design: Now in Production,* is titled "The Digital Revolution—and Beyond," but what that beyond might be does not become clear. Meggs and Purvis see digital technologies as another set of tools, not as a new model of creation, production and use that is transforming the roles of those involved in the information chain, which is how I see it. Tellingly, the last two paragraphs of the book's epilogue on page 572 read: "As so often in the past, the tools of design are changing with the advance of technology. The essence of graphic design however, remains unchanged. That essence is to give order to information, form to ideas, and expression and feeling to artifacts that document human experience. The new generation of graphic designers must take it upon themselves to define new aesthetics in electronic media and not allow the technology to define them. In doing so, they can lead the way to new and more effective approaches to their field."
37   Blauvelt and Lupton, *Graphic Design: Now in Production*.
38   Drucker and McVarish, *Graphic Design History: A Critical Guide*.
39   Ibid., xi.

East

Poster Klutsis [1,2,3,4,5]   Poster ...Y
Editorial ...er Rodchenko [1,2,3,4,5]
Poster ...ir Mayakovsky [1,3,4]
Editorial Huszar [1,3,5]   Poster ...erg Brothers [1,2,3,4,5]
Artwork Malevich [2,3,4,5]   Editorial Brodovitch [1,2,3,4,5]
Editorial Moholy-Nagy [1,2,3,4,5]
Editorial ...ky [1,2,3,4,5]   Poster...
Poster Cieslewicz [3,4,5]
Editorial ...b Soffici [2,3,5]   Editorial ...ge [3,4,5]   Ja... Poster ...ca [3,...
Editorial Marinetti [1,2,3,4,5]   Editorial Sutnar [1,3,4,5]
Poster ...tto Cappiello [2,3,4,5]   Poster Grignani [1,3,4]
Artwork ...o Balla [2,4,5]   Editorial ...to Depero [1,2,3,4,5]   Poster ...ni Pintori [3,4,5]
Editorial ...ffmann [2,3,4]
Poster Roller [1,2,3,4,5]
Poster ...Klimt [1,2,3,5]   Editorial Bayer [1,2,3,4,5]
Poster Klinger [2,3,4,5]   Artwork Höch [1,2,4,5]   Infographics ... [1,3,5]
Poster ...an Moser [1,2,3,4,5]   Editorial ...ll [2,4,5]   Typeface ...ner [2,3,4,5]
Poster Rudi Erdt [1,2,3,5]   Poster ...Schmidt [1,2,3,4,5]
Poster Behrens [1,2,3,4,5]   Editorial ...eininger [2,4,5]
...ns Heine [3,4,5]   Typeface ...ch [3,4,5]   Editorial ...witters [1,2,3,4,5]   Infographics [3,5]   Identity ...er [2,3,...
...rial ...mann [1,2,3,5]   Editorial ...artfield [1,2,3,4,5]   Identity Stankowski [1,2,3,4,5]   Typeface ...utiger [1,2,3,...
Poster ...g Hohlwein [1,2,3,4,5]   Typeface ...rs [2,3,4]   Poster H...
Poster Bernhard [1,2,3,4,5]   Editorial ...ichold [1,2,3,4,5]   Editoria...
Poster ...Keller [3,4,5]   Poster Müller-Brockmann [1,2,3,4,5]   Type...
Poster ...s Stoecklin [3,4,5]   Editorial ...uburg [2,3,5]   Editorial ...varelli [1,2,3,5]
Editorial Matter [1,2,3,4,5]   Editorial
Editorial ...s Itten [2,3,4,5]   Poster ...uber [3,4,5]   Poster ...t Leupin [3,4,5]
Poster ...aumberger [3,4,5]   Poster ...ll [1,2,3,4,5]   Editorial Paul Lohse [2,3,4]   Ed...
Editorial ...rt [1,2,3,4,5]   Typ...
Identity ...an de Velde [1,2,3,5]   Artwork ...ndrian [2,4,5]   Editorial ...huitema [1,2,3,4,5]
...itorial ...e Mallarmé [3,4,5]   Editorial ...n Doesburg [1,2,3,4,5]
...Alexandre Steinlen [2,3,4,5]   Poster ...ndre [1,2]   Poster ...olin [3,4,5]   Poster ...nd Savignac [3,...
...use-Lautrec [2,3,4,5]   Editorial ...e Apollinaire [2,3,4,5]   Poster ...arlu [1,2,3,4,5]
...ardsley [1,2,3,5]
Editorial ...rane [1,3,5]   Typeface ...ohnston [1,2,3,4,5]   Typeface ...Morison [2,3,5]
...s Rennie Mackintosh [2,3,4,5]   Poster ...Leete [2,3,4,5]   Typeface ...lt [1,2,3,4,5]   Poster Games [3,4,5]   Editori...
...Macdonald [1,2,3]   Poster Lumley [2,3,5]
...rris [1,2,3,4,5]   Artwork ...casso [2,4,5]   Editorial ...y Thompson [1,3,4,5]   Identity...
Editorial ...in [1,3,5]   Identity ...Wolf [3,...
Artwork ... [2,4,5]   Identity ...nd [1,2,3,4,5]   Editorial ...rederico [1...
Identity Golden [1,2,3,5]
Poster Montgomery Flagg [1,2,3,4,5]   Editorial ...stig [1,2,3,4,5]   E...
Poster ...n Rockwell [1,2,3]   Editorial ...eall [1,2,3,4,5]   Editorial ...eless [1,2,5]   Edi...
Identity ...Coiner [2,3,4]
Poster ...nni [3,4,5]

West

1900   1910   1920   1930   1940   1950

This overview shows the type of graphic design object with which the most frequently depicted designers (see 2.3) have been included in the five selected graphic design histories. These objects together show what the output of graphic design is according to the selected graphic design histories.

nashiro [3,4,5]

Poster ...ri Yokoo [1,3,5]

har Swierzy [3,4,5]

Identity ...o Vignelli [1,3,5]

Poster Brandt [1,2,3]
Typeface Zapf [1,4,5]
Editorial ...ckhaus [1,3,4,5]   Poster ...oesch [3,4,5]   Typeface ...ermann [2,4,5]
...ann [1,2,3,4,5]
...dermatt [1,3,4,5]
...ffmann [1,2,5]
Poster ...ng Weingart [1,2,3,4,5]
...er [2,3,4]
Poster ...uder [2,4,5]   Poster Schraivogel [3,4,5]
...rial Tissi [3,4,5]   Poster ...umbar [2,3,5]   Poster Dumbar [1,3,4,5]
...ce ...inger [1,2,3,4,5]   Poster ...n Toorn [2,3,5]
Poster ...rouwel [2,3,4,5]

Editorial Massin [2,3,4,5]   Poster ...s [2,3,4,5]

Typeface ...ody [1,2,3,4,5]
Identity ...tcher [2,3,4,5]   Identity ...eid [1,2,3,4]   Poster Sagmeister [1,2,4,5]
...shaug [3,4,5]   Typeface Carter [1,2,3,4,5]   Editorial ...Associates [3,4,5]
Editorial ...ooper [1,3,5]   Editorial ...arson [1,2,3,4,5]
...3,4,5   Editorial ...sman [3,4,5]   Editorial ...nderLans [1,2,3,4,5]
...itorial Lois [1,3,5]   Editorial ...dman [1,2,5]   Typeface ...cko [1,2,4,5]   Typeface ...edy [1,2,3]
Poster Moscoso [1,2,3,5]   Poster ...d Fella [1,2,3,4,5]
Poster ...a Kruger [2,3,4]   Typeface ...s [1,2,4]
...ial Chwast [2,3,4,5]   Poster ...reiman [1,2,3,4,5]   Poster ...ine McCoy [1,2,3,5]
...balin [1,2,3,4,5]   Poster ...Max [1,4,5]   Poster ...cher [1,2,3,5]   Editorial ...lman [1,2,3]
Poster Glaser [1,2,4,5]   Identity ...are [1,4,5]
Poster ...ilson [1,2,4,5]   Editorial ...au [1,2,3]   Digital ...aeda [2,4,5]

1970   1980   1990   2000   2010   2020

1   Drucker & McVarish, *Graphic Design History: A Critical Guide*, 2nd edition, 2013
2   Eskilson, *Graphic Design: A History*, 2nd edition, 2012
3   Hollis, *Graphic Design: A Concise History*, 2001
4   Jubert, *Typography and Graphic Design: From Antiquity to the Present*, 2006
5   Meggs & Purvis, *Meggs' History of Graphic Design*, 5th edition, 2012

in a shorter time and at a larger scale. Even more significant is the space between the thresholds. The lines of mechanization and digitization are much more spaced out than those between digitization and dissemination; to me this indicates that they succeeded each other more rapidly.

The practices that will be the subject of the three case studies in this book will be mainly situated in the sphere on the right of the dissemination threshold. My own practice is predominantly positioned between the thresholds of digitization and dissemination. Between the technologies that gave me access to the field, and those that have opened it up even further to the point that it ceases to be a specialist activity.

User, Editor, Designer

I have a graphic design practice but I was never trained as a graphic designer. I studied architectural design in the early 1990s. It was the time when the computer became the universal tool for many creative fields. Formats from one design domain became accessible to designers from other fields.

In my case, I took a bit of a detour from architectural to graphic design. In the final year of my studies I became fascinated by interactive media: the opportunities to include animation, photography and sound, the directness of screen-based production, "what you see is what you get," and the possibilities that interactivity offered beyond a single linear narrative. In one of my graduation projects I used this new technology to explore novel ways of presenting architectural ideas. For a few years after my graduation I had two practices, one focused on what I was trained to do, architectural design, and one aimed at the representation of architecture and urbanism through interactive media.

With the computer as universal tool it is difficult to link a producer to a discipline based on her output alone. That might explain why one day I was approached to do the graphic design of a book by a team of urban designers who were familiar with my interactive media projects. I enjoyed making the book so much that I decided to shift the focus of my practices towards book design at first, later expanding it to other aspects of graphic design. While making the book, I became aware that I was designing something that I actually knew as a user, something I had never noticed while working on screen-based projects. I liked designing interactive media a lot, but never actually sat down after work and put on a CD-ROM to explore it. The making of the book exposed me as a nonuser of interactive media; it felt hypocritical to continue working on these kinds of projects. I have been, since an early age, an avid reader, lover and collector of books. So when I started designing books I brought with me an understanding of the typology.

Andrew Blauvelt argues that the graphic design discipline controls the production of knowledge by defining its own languages, such as technical jargon.[40] I agree with Blauvelt that the language of a field can function as a kind of disciplinary gatekeeping. I experienced this first-hand in the transition from one domain to another. In the daily routine of my practice I noticed it in discussions with publishers and printers, but also when I was studying to learn more about the field by reading texts or listening to lectures. Over time I became familiar with the vocabulary of the field, but in those early "unlettered" days I relied on my understanding of graphic formats like certain book types and architectural drawings. My design practice was built on the tacit knowledge of typologies I had developed as a user, a sense of disciplinarity followed later.

Building further on Blauvelt's notion of the control of knowledge by a field, I think that an expanded notion of graphic design will need to address the terminology used to describe it. In the introduction I already raised this issue. In this book I have chosen the strategy of using two different "languages," a textual and a visual one, presented in parallel, to both question the singular representation of research through text alone, as well as to open up the research by offering alternatives that might be more accessible to some readers.

When I first started working in the field of graphic design, out of a lack of confidence I opted for a description of my activities rather than a disciplinary label: "I make books" or "I design books," but not "I am a graphic designer." For me, the transition to calling myself a designer was the result of recognition I received from the field. As soon as my projects were selected in competitions, were exhibited and published, I dared calling myself a graphic designer. Names that describe someone's position function as a form of gatekeeping. In this way language and titles limit a full understanding of the transformations of a field, as it prevents seeing the full spectrum of what is being made and who is making it. In this book I will use the terms "editor of graphic information" and "graphic designer" without attaching a hierarchical distinction to them. I will use "designer" as a general label for practitioners of the field and those who name themselves as such. In other cases I will use "editor."

In my work, the educational background in a discipline other than graphic design became apparent in how I processed certain kinds of content, such as architectural drawings and maps. To me these are not fixed visual sources, but documents that need to be edited to enhance their representation. Both technically, in terms of optimizing the reproduction in print or on the screen, and editorially, in terms of improving the communication of the intentions of the author of an architectural drawing or map. In other words, I redesign the map or drawing: I adjust line weights, colors, crop and scale. In some cases I reconsider the elements it consists of: making some parts less important, or taking them out altogether, while emphasizing other elements, sometimes adding data, if I think it is necessary for its understanding.

Here it is important to once more stress the impact of digital technology on the production of visual information. Above I describe the editing and redesigning of maps. Until the digital age it was not possible to do this so easily. Take, for instance, the *World Geo-Graphic Atlas*, published by the Container Corporation of America in 1953 and designed and edited by former Bauhaus tutor Austrian-born American graphic designer Herbert Bayer (1900–1985), listed in many anthologies of graphic design as a design classic.[41] For five years Bayer and his collaborators worked on this book, exploring new cartographic projection types like Richard Buckminster Fuller's Dymaxion map, and the visual statistic picture language Isotype by Otto Neurath, Marie Neurath and Gerd Arntz.[42] Yet the majority of the maps in the *World Geo-Graphic Atlas* are existing plans sourced from other atlases, reproduced without change. Here is an atlas initiated by a wealthy, powerful company, designed by a proficient designer who was given editorial control and a generous amount of time and resources to make an exceptional book. And yet

40 Blauvelt, "Graphic Design: Discipline, Medium, Practice, Tool, or Other?"
41 Bayer, *World Geo-Graphic Atlas: A Composite of Man's Environment*.
42 Lommen, *Het boek van het gedrukte boek: Een visuele geschiedenis*, 362.

L|
P|
P|
Inkjet printing [2]
Phototypesetting [2]
Varigraph lettering devic‹

Phone answering machine [2]
Teletypesetter [2]
Dot matrix printer [2]
Portable Leica camera [2]
Instamatic camera [2]
Kodachrome film [2]
Scanner (Belinograph) [2]
(Industrial) screen printing [1,2]
Process color printing [2]
Linoleum cutting tools (Linocut) [2]
or separation [2]
casting [2]

| 1900 | 1910 | 1920 | 1930 | 1940 | 1950 |

Mechanization

Timeline of tools for creating, recording, editing, producing, distributing and accessing visual information. The technologies in this overview are collected from two books: Blauvelt & Lupton, *Graphic Design: Now in Production* (2011) and Drucker & McVarish, *Graphic Design History: A Critical Guide*, 2nd edition (2013).

The timeline distinguishes three sets of technologies: mechanization, digitization and dissemination. Mechanization refers to the technologies of the industrial production of graphic information that enabled the graphic designer to emerge as a specialist in the production of visual information. The digitization

of production tools has impacted the role of the designer, the tasks she performs and her role in the information chain. Dissemination technologies impacted the speed and expanse of the distribution of information, and the interaction and exchange of that information with others. The timeline is limited to the period 1900–2020. The starting date is related to what I see as the beginning of graphic design. The technologies are organized by date of introduction and thus not necessarily the moment they were most intensively used. The sets of technologies form three thresholds or boundaries between different spheres that each enable different practices of production and use.

Adobe DPS [2]
Adobe Creative Cloud
iPad [1,2]
iA Writer
Typekit [1]
e-pub [1]
Kindle [1,2]

iPhone	5G
Adobe Creative Suite	TikTok
Adobe Premiere Pro	DouYin
Processing [1]	Periscope
Scriptographer [1]	TencentVideo
USB flash drive [2]	WeChat
Adobe InDesign	Instagram
e-reader [1]	Kickstarter [1]
Adobe After Effects	Newspaper Club [1]
FontLab [2]	WeTransfer
PDF [2]	Sina Weibo
Wacom tablet [2]	4G
Linux [1]	GitHub
Digital stock photography [2]	Instapaper [1]
Adobe Illustrator	MagCloud [1]
Adobe Photoshop [2]	Dropbox
QuarkXPress [1,2]	Tumblr
Screen reader [1]	Blurb [1]
SGML [2]	Twitter [1]
LaserWriter desktop printer [2]	YouTube [1]
Adobe PageMaker [2]	Facebook [1]
Macintosh computer [1,2]	Flickr [1]
ThinkJet inkjet printer [2]	Beidu Tieba
3-D printer [1]	WordPress
News reader (usenet) [1]	Lulu [1]
Copyleft [1]	Google image search [2]
Digital camera [2]	Wikipedia [1]
Microsoft [2]	3G
Desktop publishing [1]	XHTML [2]
Computer generated image (CGI) programs [2]	Google [1]
Digital fonts [2]	CSS
Geographic information systems (GIS) [2]	QR code [1]
Mouse [2]	HTML [1]

Risograph [1]
Electronic paper [1]
Color photocopier [2]
Pantone (matching system) [2]
IBM Selectric with changeable type [2]
set [2]
copying [2]
lithography [2]

1970   1980   1990   2000   2010   2020

Digitization | Dissemination

1   Blauvelt & Lupton, *Graphic Design: Now in Production*, 2011
2   Drucker & McVarish, *Graphic Design History: A Critical Guide*, 2nd edition, 2013

the "look and function" of an essential part of the content has not been decided by the designer.[43]

When I started working with maps I realized there are many similarities between cartography and graphic design. The output of both is primarily graphical, of course, and the tools to create, edit, produce and distribute overlap. On a more fundamental level, digital technologies have opened up both fields. The impact of new technologies was bigger and more clearly discernible in cartography than in graphic design. The making of maps used to be a very closed off field dominated by powerful elites, such as the great map houses of the West, like De Agostini Editore (Italy), Michelin (France) and Rand McNally (USA), the state, and to a lesser extent the academic world.[44] Digital technology, GPS, new mapmaking software, "open source" collaborative tools, mobile mapping applications and geotagging opened up the field of cartography and made it possible for new mapmaking practices to emerge that started to map different subjects, in novel ways, occasionally resulting in new forms. The radical change of cartography before and after the introduction of digital technologies is even bigger than the transformation that graphic design underwent in the digital age.

Another important difference between cartography and graphic design is that the impact of new technologies on mapmaking has been more substantially theorized. I am especially interested in the post-representational reading of cartography, as it takes an expanded view of cartography that also considers the use of maps.

A Post-Representational Approach to Mapmaking

Post-representational cartography understands a map as a process rather than an object. According to this approach, maps are never fully formed and their work is never complete, they are in a constant state of becoming. A map is constantly being produced and reproduced, every time a user engages with it. According to this approach to cartography, the binary division between production and application, between producer and user no longer applies.

When approached as a process, it is interesting to consider at what moment the production of a map ends.[45] Is it when it is conceptualized, when it is embedded in other content on a page, when it is printed, when it is loaded on the screen of a digital device, when it is read, or maybe never? Similar considerations can be made when contemplating the use of maps. When is it first used? During the process of creation, the moment the mapmaker sees the whole through the fragments?

What I take from post-representational cartography is the consideration that making and using are not consecutive processes but parallel tracks. The blurring of the producer–user divide and approaching the map as process rather than a product reframe cartography as a discipline of practices, not one of representations. By recasting cartography as a broad set of spatial practices it moves beyond the narrow confines of seeing the map as a product of a specialized activity. It is my ambition to do the same with graphic design: to reframe the field and consider new practices.

As the term suggests, post-representational cartography is regarded as a subsequent phase in the thinking about mapmaking. After the Second World War, thinking about maps developed into what was later called cognitive cartography[46] and representational cartography.[47] Its premise is that the world can be objectively known and truthfully mapped. From this point of view, cartography progresses by looking at self-referential methodological questions aimed at improving how a map communicates. Ideas of cognitive psychology are used to improve map designs by carefully controlled scientific experiments. In a next chapter of this book I will go deeper into the research of French cartographer and theorist Jacques Bertin (1918–2010), one of the key representatives of this mode of thinking.

At the end of the 1980s, ideas from poststructuralist thinking, social constructionism and actor-network theory resulted in a shift in the thinking about maps. In what is called critical cartography[48] and more-than-representational cartography[49] maps are seen as the products of power, but also as producing power themselves. Mapmaking is not a neutral, objective pursuit, but one laden with power. It deals with creating knowledge, rather than with revealing it.

The early 2000s saw a third shift in the thinking about maps: post-representational cartography. Representational cartography assumed that the world was knowable and mappable and did not question the nature of the map itself. Neither is the map fundamentally questioned in more-than-representational cartography. It might regard the map as diverse, rhetorical, relational and complex, but nonetheless as a stable product: a map. In post-representational cartography, however, the fundamental status of the map is questioned. Maps are now understood as never fully formed, the work on them never complete. Rob Kitchin, professor of geography at the National University of Ireland Maynooth, and Martin Dodge, senior lecturer in human geography at the University of Manchester, use the terms ontological and ontogenic to describe the difference between post-representational thinking about mapmaking and the two previous approaches.[50] Ontological refers to how things are, ontogenic to how things become. Kitchin and Dodge describe representational cartography and more-than-representational cartography as having an ontologically secure foundation: a map is a map and it is constant. In post-representational cartography, however, the fundamental question of cartography is not ontological but ontogenic, how does a map become.[51] Unlike the cartographic theories that preceded it, post-representational cartography looks at the full process of mapmaking, from conception to use, from producer to user.

The three different approaches to cartography described above—representational, more-than-representational and post-representational cartography—are all based on distinct epistemological methods. Representational cartography uses quantitative methods from cognitive psychology. In more-than-representational cartography methods of textual and linguistic deconstruction are predominantly employed. Post-representational cartography's processual approach employs a variety of methods, including genealogies, ethnographies, ethnomethodology, participant observation, observant participation and deconstruction, to open up or entangle the practices of mapmaking. This last set of methods is able to capture the full scope of how a map is created and produced, to how it is reproduced every time a user engages with it and how social, embodied, political and economic relations play a role in this messy process.

Sébastien Caquard, associate professor in geography, environment and urban planning at Concordia University in Montréal, has argued that the supposed separation between critical theory of more-than-representational cartography and

43 *Oxford English Dictionary*, "design," accessed 5 October 2018, https://en.oxford-dictionaries.com/definition/design.
44 Crampton and Krygier, "An Introduction to Critical Cartography," 12.
45 Del Casino Jr. and Hanna, "Beyond The 'Binaries': A Methodological Intervention for Interrogating Maps as Representational Practices," 35.
46 Caquard, "A Post-Representational Perspective on Cognitive Cartography."
47 Kitchin, "The Transformation of Cartographic Thought."
48 Crampton and Krygier, "An Introduction to Critical Cartography."
49 Kitchin, "The Transformation of Cartographic Thought."
50 Kitchin and Dodge, "Rethinking Maps."
51 Ibid., 5.

East

Gustav Klutsis [1,2,3,4,5]

Ryuichi Y

Alexander Rodchenko [1,2,3,4,5]

Vladimir Mayakovsky [1,3,4]

Vilmos Huszar [1,3,5]　Stenberg Brothers [1,2,3,4,5]

Kasimir Malevich [2,3,4,5]　Alexey Brodovitch [1,2,3,4,5]

László Moholy-Nagy [1,2,3,4,5]

El Lissitzky [1,2,3,4,5]

Wald

Roman Cieslewicz [3,4,5]

Ardengo Soffici [2,3,5]　Karel Teige [3,4,5]

Jan Lenica [3,

Filippo Marinetti [1,2,3,4,5]　Ladislav Sutnar [1,3,4,5]

Leonetto Cappiello [2,3,4,5]

Franco Grignani [1,3,

Giacomo Balla [2,4,5]　Fortunato Depero [1,2,3,4,5]

Giovanni Pintori [3,4,5]

Josef Hoffmann [2,3,4]

Alfred Roller [1,2,3,4,5]

Gustav Klimt [1,2,3,5]

Herbert Bayer [1,2,3,4,5]

Julius Klinger [2,3,4,5]　Hannah Höch [1,2,4,5]　Otto Neurath [1,3,5]

Koloman Moser [1,2,3,4,5]　Hugo Ball [2,4,5]　Paul Renner [2,3,4,5]

Hans Rudi Erdt [1,2,3,5]　Joost Schmidt [1,2,3,4,5]

Peter Behrens [1,2,3,4,5]　Lyonel Feininger [2,4,5]

nas Heine [3,4,5]　Rudolf Koch [3,4,5]　Kurt Schwitters [1,2,3,4,5]　Gerd Arntz [1,3,5]

Otl Aicher [2,3,

Eckmann [1,2,3,5]　John Heartfield [1,2,3,4,5]　Anton Stankowski [1,2,3,4,5]

Adrian Frutiger [1,2,3,

Ludwig Hohlwein [1,2,3,4,5]　Josef Albers [2,3,4]

Armin H

Lucian Bernhard [1,2,3,4,5]　Jan Tschichold [1,2,3,4,5]

Siegfrie

Ernst Keller [3,4,5]　Josef Müller-Brockmann [1,2,3,4,5]

Edua

Niklaus Stoecklin [3,4,5]　Hans Neuburg [2,3,5]　Carlo Vivarelli [1,2,3,5]

Herbert Matter [1,2,3,4,5]

Karl Ger

Johannes Itten [2,3,4,5]　Max Huber [3,4,5]　Herbert Leupin [3,4,5]

Otto Baumberger [3,4,5]　Max Bill [1,2,3,4,5]　Richard Paul Lohse [2,3,4]　L

Piet Zwart [1,2,3,4,5]　Ph

Henry van de Velde [1,2,3,5]　Piet Mondrian [2,4,5]　Paul Schuitema [1,2,3,4,5]　Ph

éphane Mallarmé [3,4,5]　Theo van Doesburg [1,2,3,4,5]　Inkjet printing [2]

Alexandre Steinlen [2,3,4,5]　Cassandre [1,2]　Paul Colin [3,4,5]　R　Phototypesetting [2]

use-Lautrec [2,3,4,5]　Guillaume Apollinaire [2,3,4,5]　Jean Carlu [1,2,3,4,5]　Varigraph lettering devic

ardsley [1,2,3,5]　Phone answering machine [2]

Walter Crane [1,3,5]　Edward Johnston [1,2,3,4,5]　Teletypesetter [2]　rison [2,3,5]

les Rennie Mackintosh [2,3,4,5]　Alfred Leete [2,3,4,5]　Dot matrix printer [2]　Abram Games [3,4,5]　Antho

Macdonald [1,2,3]　Saville Lumley [2,3,5]　Portable Leica camera [2]

rris [1,2,3,4,5]　Pablo Picasso [2,4,5]　Instamatic camera [2]　Bradbury Thompson [1,3,4,5]　Saul Ba

Kodachrome film [2]　Will Burtin [1,3,5]　Henry Wolf [

Scanner (Belinograph) [2]　n Ray [2,4,5]　Paul Rand [1,2,3,4,5]　Gene Frederico [

(Industrial) screen printing [1,2]　William Golden [1,2,3,5]

Process color printing [2]　James Montgomery Flagg [1,2,3,4,5]　Alvin Lustig [1,2,3,4,5]　S

Linoleum cutting tools (Linocut) [2]　Rockwell [1,2,3]　Lester Beall [1,2,3,4,5]　Cipe Pineless [1,2,5]　He

or separation [2]　Charles Coiner [2,3,4]

casting [2]　Leo Lionni [3,4,5]

1900　　　1910　　　1920　　　1930　　　1940　　　1950

Mechanization

Comparison between the most frequently depicted designers in the five selected
design histories (2.3) and the model of technological thresholds (2.5).

Tadanori Yokoo [1,3,5]

ashiro [3,4,5]

ar Swierzy [3,4,5]

Massimo Vignelli [1,3,5]

Adobe DPS [2]
Adobe Creative Cloud
iPad [1,2]
iA Writer
Typekit [1]
e-pub [1]
Kindle [1,2]
iPhone   5G
Adobe Creative Suite   TikTok
Adobe Premiere Pro   DouYin
Processing [1]   Periscope
Scriptographer [1]   TencentVideo
USB flash drive [2]   WeChat
Adobe InDesign   Instagram
e-reader [1]   Kickstarter [1]
Adobe After Effects   Newspaper Club [1]
FontLab [2]   WeTransfer
PDF [2]   Sina Weibo
Wacom tablet [2]   4G
Linux [1]   GitHub
Digital stock photography [2]   Instapaper [1]
Adobe Illustrator   MagCloud [1]
Adobe Photoshop [2]   Dropbox
QuarkXPress [1,2]   Tumblr
Screen reader [1]   Blurb [1]
SGML [2]   Twitter [1]
LaserWriter desktop printer [2]   YouTube [1]
Adobe PageMaker [2]   Facebook [1]
Macintosh computer [1,2]   Flickr [1]
ThinkJet inkjet printer [2]   Beidu Tieba
3-D printer [1]   WordPress
News reader (usenet) [1]   Lulu [1]
Copyleft [1]   Google image search [2]
Digital camera [2]   Wikipedia [1]
Microsoft [2]   3G
Desktop publishing [1]   XHTML [2]
Computer generated image (CGI) programs [2]   Google [1]
Digital fonts [2]   CSS
Geographic information systems (GIS) [2]   QR code [1]
Mouse [2]   HTML [1]

Peter Brandt [1,2,3]
Hermann Zapf [1,4,5]
Willy Fleckhaus [1,3,4,5]   Uwe Loesch [3,4,5]   Erik
ann [1,2,3,4,5]
dermatt [1,3,4,5]
offmann [1,2,5]
Wolfgang Weingart [1,2]   Color photocopier [2]
Pantone (matching system) [2]
IBM Selectric with changeable type [2]
set [2] Tissi [3,4,5]   Gert Dumbar [2,3,5]
copying [2,3,4,5]   Jan van Toorn [2,3,5]
ithography [2] Wim Crouwel [2,3,4,5]
Robert Massin [2,3,4,5]   Grapus [2,3,4,5]

Risograph [1]
Electronic paper [1]

Alan Fletcher [2,3,4,5]   Jamie Reid [1,2,3,4]   Stefan Sagmei
shaug [3,4,5]   Matthew Ca... Why Not Associates [3,4,5]
Muriel Cooper [1,3,5]   David Carson [1,2,3,4,5]
Lou Dorfsman [3,4,5]   Rudy VanderLans [1,2,3,4,5]
orge Lois [1,3,5]   Dan Fried... ...na Licko [1,2,4,5]   Jeffery Keedy [1,2,3]
Victor Moscoso   Edward Fella [1,2,3,4,5]
Barbara Kruger [2,3,4]   Elliott Earl
ur Chwast [2,3,4...]   Katherine McCoy
balin [1,2,3,4,5]   Pe... Paula Scher [1,2,3,5]   Tibor Kalm...
Milton G... ...Kare [1,4,5]
Wes...   Bruce Mau [1,2,3]   John Maeda [2,4,5]

1970	1980	1990	2000	2010	2020

Digitization   Dissemination

1   Drucker & McVarish, *Graphic Design History: A Critical Guide*, 2nd edition, 2013
2   Eskilson, *Graphic Design: A History*, 2nd edition, 2012
3   Hollis, *Graphic Design: A Concise History*, 2001
4   Jubert, *Typography and Graphic Design: From Antiquity to the Present*, 2006
5   Meggs & Purvis, *Meggs' History of Graphic Design*, 5th edition, 2012
1   Blauvelt & Lupton, *Graphic Design: Now in Production*, 2011
2   Drucker & McVarish, *Graphic Design History: A Critical Guide*, 2nd edition, 2013

empiricist practice of representational cartography might be resolved by focusing on mapmaking as a process.[52] To Caquard, post-representational cartography offers opportunities to combine these two distinct approaches to strengthen the understanding of the mental, emotional and embodied relationships with maps, and with places through maps.[53]

A similar empiricist/critical disjuncture can be observed in a subfield of graphic design called information design. In a recent book on the domain, it is defined as clarifying complex information with the needs of users in mind.[54] Testing is seen as a key part of that process. Various iterative methods from behavioral research are employed to create designs that fit the needs of people in a specific context. Information design practices have little in common in terms of approach, methods and criteria, with those included in the aforementioned *Graphic Design: Now in Production*.

Blind Map

There is a type of map that does not contain any text labels. It is mainly used in education, intended for pupils to fill in the missing names in tests. In the English language it is called an unlettered map. In Swedish it is a map without names: *namnlös karta*.[55] In some languages it is described as mute: *carte muette* in French, *carta muta* in Italian, *carta muda* in Portuguese, *mapa mudo* in Spanish, *stumme Karte* in German. Other languages designate it blind: *blinde kaart* in Dutch, *blindkart* in Norwegian, *blindkort* in Danish. Whatever adjective used, all highlight that something is missing and that it is up to the user to complete the map. Of all the map types I know this is the one that most clearly shows its fundamental emergent status. I will use the term *blind map* for any graphically produced object to highlight its fundamental emergent status. Any map, any piece of graphic design is a blind map as it is never fully formed, but completed every time a user engages with it.

The blind map as transdisciplinary concept emphasizes once more the similarities between cartography and graphic design. The fields may have different origins that resulted in disparate educational structures, different criteria to evaluate the output, and distinct theoretical frameworks, however, the current practices of both disciplines are practically the same. They use the same or related tools to record, create, edit, produce and distribute the visual information they produce. As someone whose work is positioned in both fields I feel the disciplines on a practical level are merging and are becoming interchangeable.

Put differently, technological developments make the disciplines of cartography and graphic design become less distinguishable as they dissolve in the larger field of graphic representation. That is the perception of French geographer Denis Eckert, research director of the Franco-German Research Centre for the Social Sciences Centre Marc Bloch in Berlin. At a recent conference in Berlin where Eckert lectured about innovation in cartography, I asked him if he sees a difference between cartography and graphic design.[56] Eckert initially joked that cartographers tend to look down on graphic designers because they feel threatened that others, noncartographers and especially graphic designers who are skilled users of design tools, are making maps. He continued by admitting that it is difficult to uphold a strict difference between the two fields. Cartography and graphic design should be seen as subsets of a wider field of representation and communication. Eckert went on to say that there is a difference between the two in that cartographers have a familiarity with specific cartographic considerations like the ones developed and formulated in the 1960s–1980s by the aforementioned Jacques Bertin and others.

I follow Eckert's reasoning, but I want to add that the opposite is also true. Graphic design has a certain specialist knowledge that distinguishes itself from cartography, such as theories on topics like typography, visual rhetoric and grid systems. Like Eckert, I see the two fields merging or dissolving into a wider field of graphic representation. Unlike Eckert, I would not describe this wider field as representation and communication or as visual communication. Dutch publisher and design writer Hugues Boekraad has argued that communication has become essential for all design disciplines and cannot be regarded as a competency exclusive to graphic design and other disciplines aimed at creating visual symbols.[57] To Boekraad visual communication is also a troublesome term to cover the activities of graphic design and related fields because of the general dominance of the visual nowadays. I agree with Boekraad, and describe the wider discipline that includes graphic design and cartography as the field of graphic representation.

My research project positions itself in this wider field of graphic representation, applying theories from both graphic design and cartography. It is my ambition, through this book, to contribute to the consolidation of this expanded discipline. The research also aims for this wider field to include practices that are not educated or specialized in one of the two disciplines, encompassing practices from outside the fields that are using the tools of graphic design and cartography, the technologies to record, create, edit, produce and distribute visual information.

Besides distinguishing specialist knowledge, to me the main difference between cartography and graphic design is the impact of digital technology on both fields. Evolving tools and platforms transformed both disciplines. But as the field of mapmaking was more closed off, the impact of technological developments on cartography seems more substantial and more clearly discernible. As someone who works in both fields, I feel that both disciplines have been roughly awakened by the introduction of digital technologies, but where some parts of graphic design are still in a drowsy state of denial, cartography is already wide awake. This makes mapmaking a suitable subject to study and understand the wider field of graphic representation and how it is developing. I will research practices that have appropriated the tools of cartography and graphic design to make maps. Employing a mix of critical and empirical approaches, and using a variety of languages both textual and visual, the research will focus on the processes of these practices to better understand the ever-evolving field of graphic representation.

The mapmaking practices that will be investigated in the following chapters embody various aspects of the processual character of maps. The three case studies explore how design, production and distribution play a role in how a map "becomes," in the post-representational sense of the word. The case study of the Blue Dot argues that Google Maps is a processual map because the user is coproducing it. This is supported by the design of the map: the pale colored map

52 Caquard, "A Post-Representational Perspective on Cognitive Cartography," 226.
53 Ibid., 232.
54 Black, Luna, Lund, and Walker, *Information Design: Research and Practice*, xi.
55 *Enzyklopädisches Wörterbuch Kartographie in 25 Sprachen*, "unlettered map."
56 Eckert, "Is Innovation in Cartography a Mere Illusion?"
57 Boekraad, "Graphic Design as Visual Rhetoric: Principles for Design Education," 6.

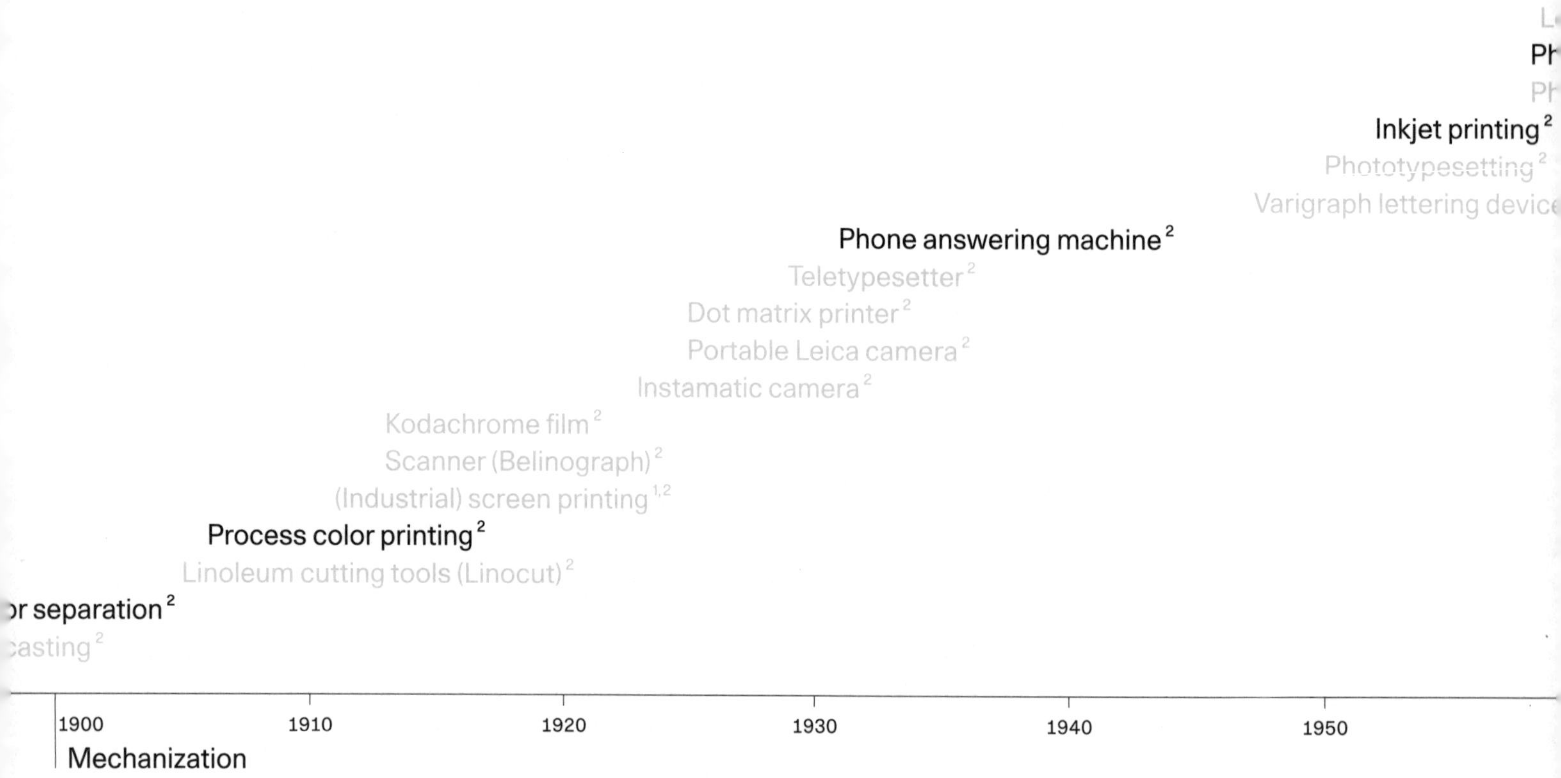

L

Ph

Ph

Inkjet printing ²

Phototypesetting ²

Varigraph lettering device

Phone answering machine ²

Teletypesetter ²

Dot matrix printer ²

Portable Leica camera ²

Instamatic camera ²

Kodachrome film ²

Scanner (Belinograph) ²

(Industrial) screen printing ¹,²

Process color printing ²

Linoleum cutting tools (Linocut) ²

or separation ²

casting ²

| 1900 | 1910 | 1920 | 1930 | 1940 | 1950 |

Mechanization

Comparison between the model of technological thresholds (2.5) and the
technologies used by graphic design studio SJG.

Adobe DPS [2]

Adobe Creative Cloud
iPad [1,2]
iA Writer
Typekit [1]
e-pub [1]
Kindle [1,2]

iPhone   5G
Adobe Creative Suite   TikTok
Adobe Premiere Pro   DouYin
Processing [1]   Periscope
Scriptographer [1]   TencentVideo
USB flash drive [2]   WeChat
Adobe InDesign   Instagram
e-reader [1]   Kickstarter [1]
Adobe After Effects   Newspaper Club [1]
FontLab [2]   WeTransfer
PDF [2]   Sina Weibo

Risograph [1]
Electronic paper [1]   Wacom tablet [2]   4G
Color photocopier [2]   Linux [1]   GitHub
Pantone (matching system) [2]   Digital stock photography [2]   Instapaper [1]
IBM Selectric with changeable type [2]   Adobe Illustrator   MagCloud [1]
set [2]   Adobe Photoshop [2]   Dropbox
copying [2]   QuarkXPress [1,2]   Tumblr
ithography [2]   Screen reader [1]   Blurb [1]
SGML [2]   Twitter [1]
LaserWriter desktop printer [2]   YouTube [1]
Adobe PageMaker [2]   Facebook [1]
Macintosh computer [1,2]   Flickr [1]
ThinkJet inkjet printer [2]   Beidu Tieba
3-D printer [1]   WordPress
News reader (usenet) [1]   Lulu [1]
Copyleft [1]   Google image search [2]
Digital camera [2]   Wikipedia [1]
Microsoft [2]   3G
Desktop publishing [1]   XHTML [2]
Computer generated image (CGI) programs [2]   Google [1]
Digital fonts [2]   CSS
Geographic information systems (GIS) [2]   QR code [1]
Mouse [2]   HTML [1]

| 1970 | 1980 | 1990 | 2000 | 2010 | 2020 |

Digitization   Dissemination

1   Blauvelt & Lupton, *Graphic Design: Now in Production*, 2011
2   Drucker & McVarish, *Graphic Design History: A Critical Guide*, 2nd edition, 2013

looks empty and thus invites the user to participate in the production process. The second case study looks into the deceivingly familiar visual language of the design of the Strava Global Heatmap and demonstrates how this supposed familiarity prevents its understanding. The third case study looks into the practices of amateur conflict mapmakers and, more specifically, how their distribution strategies utilize social media platforms. I argue that the maps of these nonspecialists have a high level of accountability because their work is embedded in a public debate.

Conclusion

I began this chapter by defining a design as the product of a process. In post-representational cartography, a map is regarded as a process rather than a product. Graphic design is processual in a similar way. The full process of creating, editing, producing, distributing and using graphical information is never finished. Every time a user engages with a design it is reproduced again and again. Rather than defining it as a product, to me the output of graphic design is a process, a blind map: unfinished and in need of a user to complete it.

It is difficult to uphold a strict division between producer and user in this extended view of the field. Not only because graphic design is processual, but also because it has ceased to be an exclusive specialist domain. Digital technologies have opened up and exposed the enigmatic and invisible processes of creating graphic design. The designer as someone with a specific training or specialized professional practice is no longer the sole producer of graphic design.

The transformation of the field of graphic design in the digital age resulted in many cases either in the denial of its practitioners, that is, graphic designers adopting outdated technologies such as letterpress, them becoming the developers of new design tools, like the design of typefaces, or appropriating other roles, such as those documented in *Graphic Design: Now in Production*. One of those roles could be the documentation, description and theorization of the methods and output of the editors of graphic information coming from outside the field that have appropriated the tools of graphic design. This learning from, to quote Venturi, Scott Brown and Izenour's *Learning from Las Vegas* (1972),[58] could bridge the gap between the traditional field and the new players. It is my ambition to reshape my design practice into a research and design practice that does exactly this. I deliberately write design and research because the description of the work of the new players in text alone is not enough. Text, and specifically disciplinary terminology, needs to be questioned, alternative formats need to be developed and other "languages" need to be designed.

Before the digital age, graphic design as a discipline was not only closed off because it required specific skills and knowledge of certain tools and technologies to enter. Specialist language employed by its practitioners, like technical jargon, also functioned as a form of gatekeeping. While the tools of graphic design democratized, its terminology did not. A study of the continuous transformation of graphic design should therefore also address the languages used in its processes. One of the aims of this research is to develop new languages, open to outsiders of the field, to describe graphic design. For this reason the book uses two different languages, a textual and a visual. In the previous chapter on concepts and methods the reasoning behind this choice is further explained.

Although different in origin, the current practices of graphic design and cartography are virtually the same. Both disciplines use similar tools and are merging to become part of the wider field of graphic representation. In this emerging field, various graphic formats like books, maps, apps, websites, information graphics and animations are produced by a group of practitioners from a variety of creative fields, but also by amateurs and commercial enterprises. As I noticed at a conference about mapping as an interdisciplinary method, this blurring of boundaries can cause unrest among some practitioners who see this as a threat. They want to hold on to their position and the accomplishments of their field. This in turn may lead to introspection and even orthodoxy about the ideas and heritage of a discipline. It is questionable if a fixed doctrine about a field can be maintained when formats, like maps, and the roles of user, producer and others in the information chain are in constant flux. And while I wonder how these two dynamic entities, of format and user/producer, might align and cause understanding of the message transferred, I believe that in this situation a certain agility to switch roles seems more appropriate than an expert understanding of a format. This in turn makes me question my role as a specialist.

Like graphic design, cartography is shaped by technological developments of the tools used in its practices. Digital technology empowered new players to enter the field who, with no prior knowledge of cartography, started mapping different subjects, in novel ways, occasionally resulting in new forms. A similar development happened in the field of graphic design. But as the field of mapmaking was very closed off, the impact of technological developments on cartography seem more substantial. The proximity of the fields, as well as the substantial larger impact digital technology had, make mapmaking a suitable subject to study the ever-evolving transformation of the field and practices of graphic design. This research will continue with a series of case studies of current mapmaking practices by technology companies and amateurs to better understand the development of the field of graphic design.

This text is an introduction to, and contextualization of, further research into the changing relationship between editors and users of graphic representation and the tools they employ to record, create, edit, produce, distribute and access visual information. The text also positions me, one-time outsider turned practitioner, as both the research subject and the one carrying it out. Indirectly, ambiguity would always be one of the topics of this study as it is undertaken in the field of artistic research that combines discursive and artistic approaches, both in its processes and products. The ambivalence of my position, due to my background and the methods I use, is something that I cannot deny and it will therefore become an essential part of my research. Ambiguity will be used as a strategy in the methods and output of this research.

58 *Learning from Las Vegas: The Forgotten Symbolism of Architectural Form* has been an important reference for this research. Both in the ambition of the authors to try to make sense as architects of a "nonarchitecture," in its attitude to "withhold judgment" as the authors write in the introduction, and in its exploration of various formats to document the research like texts, photographs, drawings, maps and diagrams.

The five researched graphic design histories are combined and split by content according to the time zones of the technological threshold model: pre-mechanization (1), mechanization (2), digitization (3) and dissemination (4).

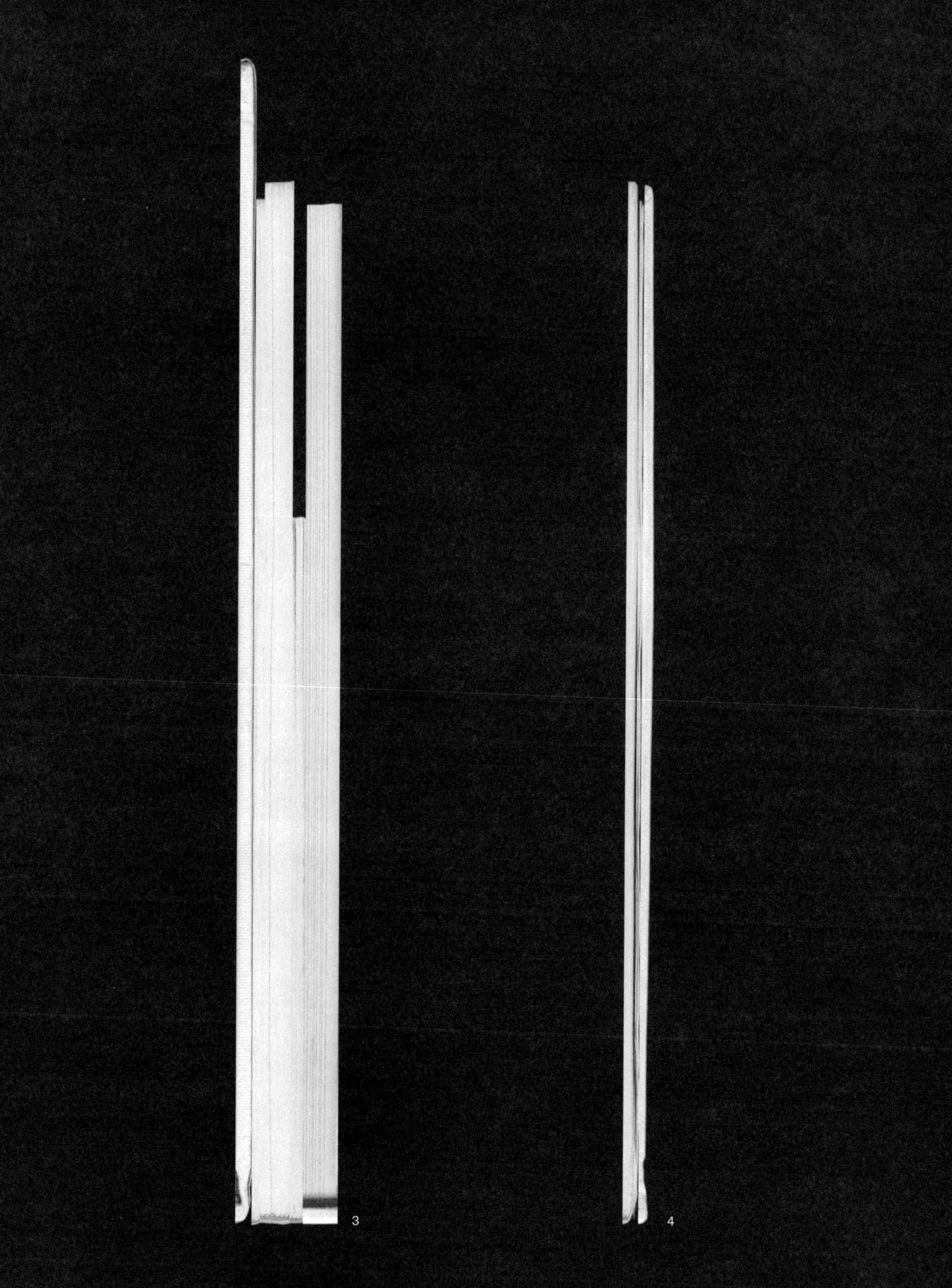

3

4

59

3   **The Blue Dot**

In this chapter I connect cartographic thinking, theories about how a map functions and its relation to the world, to design innovation, that is: to changes in established ideas about the look and functioning of a graphic product. It is my claim that the way a map is conceptually understood will determine if changes in its design are recognized as improvements. Different modes of thinking about cartography, technological advancements in the production of maps and the emergence of new practices are taken into consideration to answer the question: Is it possible to produce fundamentally new maps, and if so, how?

This text is made up of three parts that each deal with a different approach to mapmaking: representational, more-than-representational and post-representational cartography, as described in Chapter 2. These different modes of thinking are used to investigate Google Maps, the most-used map of today. First, I compare French cartographer and theorist Jacques Bertin's ideas with the design of Google Maps. Then, using ideas from critical cartography, I describe how Google Maps produces power, but is also a product of power. I describe how critical thinking informed the ambiguous design strategies developed in my design practice. Finally, taking a post-representational approach, focusing on the processual aspects of the production and use of maps, I investigate the Blue Dot in the Google Maps app. I consider its functionality, its genesis, its technological context, and explore alternative design choices.

Efficiency of Communication

In a 2018 talk entitled "Is Innovation in Cartography a Mere Illusion?,"[1] French geographer Denis Eckert stated that cartography has certain specific restraints that limit the creation of new relevant spatial visualizations.[2] This observation is striking, given the outburst of mapmaking practices that cropped up in the past decades. Since the 1990s, technological advancements have resulted in extended possibilities of combining spatial information with georeferenced databases and maps. Also, digital technologies such as smartphones and various visualization tools gave the general public access to cartography. To Eckert, however, the practices that emerged from these developments use heterogeneous spatialized information in ways that produce "maps" that are but mere superpositions of noncoordinated graphical signs without any systematic semiology. One such practice, according to Eckert, is Google Maps, the world's most-used map.

Google Earth was launched in June 2005 and came to public prominence during hurricane Katrina in August 2005, as it enabled individuals to see the dramatic changes in the landscape that had occurred. Google Maps is the cartographic overlay of Google Earth. The two programs merged into a single geospatial application that has more than 1 billion monthly users.[3] More than 1 million websites incorporate data from Google Maps.[4] According to US business news website *The Manifest*, 77 percent of all smartphone users regularly use navigation apps and of those Google Maps is the most popular by a wide margin.[5] Like most online services, Google's online cartographic platform is regularly renewed to update the information displayed and to optimize its interface. Why then is Google Maps questioned by Eckert?

1   Eckert, "Is Innovation in Cartography a Mere Illusion?" Denis Eckert is currently the research director of the Franco-German Research Centre for the Social Sciences Centre Marc Bloch in Berlin.
2   In this text I will use *visualization* as specified in the Oxford English Dictionary's first definition of the term, a "representation of an object, situation, or set of information as a chart or other image," and not as the OED's second description, "the formation of a mental image of something." For the verb *to visualize* I use the second definition in OED, to "make something visible to the eye," rather than the OED's first description, to "form a mental image of."
3   Google CEO Sundar Pichai mentions this number in his keynote address at Google's developer conference "I/O 2017." Pichai, "Google Keynote."
4   "A fresh new look for the Maps API, for all one million sites."
5   In a survey of 395 navigation app users, 67 percent preferred Google Maps, 12 percent Waze (community-driven GPS navigation app launched in 2006, originally from Israel but acquired in 2013 by Google), and 11 percent Apple Maps (mapping service launched in 2012 by Apple). Panko, "The Popularity of Google Maps: Trends in Navigation Apps in 2018."

BOUNDARIES

Equator

- - - - - - - -	ZL00–ZL03

Neighborhood

▭	ZL07–ZL13
▭	ZL14
LABEL	Neighborhood level 1 ZL07–ZL13
LABEL	Neighborhood level 2 ZL11–ZL13
LABEL	Neighborhood level 3 ZL11–ZL13

Capital city

◉	ZL00–ZL04
LABEL	ZL00–ZL09
LABEL	ZL10
LABEL	ZL11
▭	ZL01–ZL08
▭	ZL09

City/Locality

○	ZL03–ZL06
LABEL	ZL00–ZL09
LABEL	ZL10
LABEL	ZL11
LABEL	ZL12
▭	ZL01–ZL08
▭	ZL09

Region

▭	ZL00–ZL18
LABEL	ZL00
LABEL	ZL01
LABEL	ZL02
LABEL	ZL03
▭	ZL02–ZL09
▭	ZL10

Country

LABEL	ZL00–ZL06
▭	ZL00–ZL18
▭	ZL00–ZL08
▭	ZL09

Disputed

▭ ▭ ▭	ZL00–ZL18

BUILT-UP AREA

Built-up area

▬	ZL04–ZL11
▬	ZL12
▬	ZL13
▬	ZL14–ZL18

Ground (non built-up area)

▬	ZL14–ZL18

Areas of interest

▬	ZL09–ZL12
▬	ZL13
▬	ZL14–ZL18

Hospitals

▬	ZL09–ZL18

Services

LABEL	ZL12–ZL18
📍	ZL12–ZL18

Health

LABEL	ZL10–ZL18
📍	ZL10–ZL18

Place of worship

LABEL	ZL10–ZL18
📍	ZL10–ZL18

Civil services

LABEL	ZL12–ZL18
📍	ZL12–ZL18

Hotel

LABEL	ZL11–ZL18
📍	ZL11–ZL18

Food and drink

LABEL	ZL11–ZL18
📍	ZL11–ZL18

Shop

LABEL	ZL11–ZL18
📍	ZL11–ZL18

Entertainment/Leisure

LABEL	ZL11–ZL18
📍	ZL11–ZL18

Outdoor

LABEL	ZL11–ZL18
📍	ZL11–ZL18

TRAFFIC NETWORK

Airport

LABEL	ZL11–ZL18
📍	ZL11–ZL18

Public transport

LABEL	ZL09–ZL18
📍	ZL09–ZL18

Google does not offer a legend of Google Maps. The legend on this page is based on empirical research, measuring color values on different zoom levels. Google Maps contains nineteen zoom levels, 00 until 18, each with its own cartographic information, which in some cases also changes slightly in color per zoom level.

Train station

LABEL	ZL09–ZL18
📍	ZL09–ZL18

Railway

——	ZL08–ZL18

Main highway

LABEL	ZL10–ZL18
▭	ZL02–ZL18
→	ZL13–ZL18

Secondary highway

LABEL	ZL13–ZL18
▭	ZL11–ZL18
→	ZL13–ZL18

Road and street

LABEL	ZL10–ZL18
——	ZL10–ZL18
→	ZL13–ZL18

Bicycle path

——	With bicycle lane	ZL12–ZL18
- - -	Without bicycle lane	ZL12–ZL18

HYDROGRAPHY

Canal

LABEL	ZL09–ZL18
▭	ZL07–ZL18

River

LABEL	ZL03–ZL18
▭	ZL03–ZL18

Lake

LABEL	ZL06–ZL18
▭	ZL03–ZL18

Ocean

LABEL	ZL00–ZL18
▭	ZL00–ZL18

Shipping route

LABEL	ZL06–ZL18
- - -	ZL05–ZL18

TOPOGRAPHY

Vegetation

▭	Desert, almost no vegetation ZL00–ZL01
▭	ZL02
▭	Semi-desert, sparse vegetation ZL00–ZL01
▭	ZL02
▭	Moderate vegetation ZL00–ZL01
▭	Dense vegetation ZL00–ZL01
▭	Very dense vegetation ZL00–ZL01

Woodland, park and conservation area

LABEL	ZL03–ZL18
▭	ZL02
▭	ZL02
▭	ZL02
▭	ZL03–ZL18
▭	ZL03–ZL18
▭	ZL03–ZL18

Ice sheet and polar desert

	ZL00
	ZL01
▭	ZL02

Production land

	ZL03
▭	ZL04–ZL05
▭	ZL06
▭	ZL07
▭	ZL08–ZL09
▭	ZL10
▭	ZL11
▭	ZL12
▭	ZL13

Beach

LABEL	ZL12–ZL18
▭	ZL06–ZL18

Mountains

▭	ZL00–ZL10
LABEL	Mountain peak ZL05–ZL18
📍	ZL05–ZL18

PLACE RESULT

Search result

LABEL	ZL00–ZL18
📍	ZL00–ZL18

Starred

LABEL	ZL00–ZL18
📍	ZL00–ZL18

Favorite

LABEL	ZL00–ZL18
📍	ZL00–ZL18

Want to go

LABEL	ZL00–ZL18
📍	ZL00–ZL18

My Location (Blue Dot)

●	ZL00–ZL18

Denis Eckert's objection to Google Maps is that the information it provides is too heterogeneous: it is an "assemblage without design or reflection."[6] To him, a map should be an organized construction of graphical signs. According to Eckert, maps should not contain fuzzy, unnecessary information, but instead be built on a rigorously selected set of relevant data. A map then is a combination of selected data rendered in a systematic legend that shows an understandable structure that can be easily reproduced mentally. In his lecture, Eckert showed a screenshot of Google Maps that was difficult to decipher. The map contained many symbols that did not feel like part of a consistent system in terms of shape and color. Also, there seemed little coherence between the map's symbols in the foreground and the base map in the background. In short, Eckert questions the effectiveness of the communicative aspects of Google Maps.

In an earlier text that Eckert wrote with French geographer Laurent Jégou, the projection method of Google Maps 2008 is criticized.[7] The article also points out errors in the scale bars in the cartographic application. Some of the issues described in the article have since been resolved by Google, but it is interesting to look at the more fundamental critique. According to Eckert and Jégou, in Google Maps technical choices take prevalence over cartographic ones. They suspect that in developing the software the technical constraint of the display speed was more important than the relevance of the geographical representation. Google Maps and other similar applications, according to the article, are developed without prior knowledge of either the codes of mapmaking or any cartographic reflection. The authors conclude that they regret that Google Maps is not as reliable and coherent in a cartographic sense as it is rich and instantaneous in its technical capabilities.

In his 2018 lecture, Eckert defines a map as a visual representation encoded by given, stable rules and made for specific purposes. He considers this conception to be in line with a particular French approach to cartography, referring to the work of French geographer Roger Brunet, particularly his book *La carte, mode d'emploi* (1987) and most notably the work of French cartographer and theorist Jacques Bertin (1918–2010), specifically the book *Semiology of Graphics*, original title *Sémiologie graphique* (1967).

Using concepts from semiotics, the study of signs and sign processes, Bertin understood the map as a sign system. His approach can be considered structuralist as he focuses on the relationship between the elements of graphics and not on the elements themselves. Bertin's research is in line with that of other thinkers and philosophers who analyzed images in the period from the 1950s to the 1970s, most notably the work of French philosopher Roland Barthes (1915–1980) on advertisements, but also the work of French philosopher and sociologist Pierre Bourdieu (1930–2002) on photography. Bertin's goal was to improve the effectiveness of maps. To Bertin, a map is a monosemic system.[8] In such a system the unique meaning of each sign is specified by a legend. By contrast, in a polysemic system the meaning of an individual sign follows the consideration of the collection of signs as a whole. Signification thus becomes subjective and debatable. Abstract imagery represents an extreme form of polysemy. Signs do not signify anything precise and thus the image becomes pansemic. According to Bertin, all participants in the making and reading of a monosemic system, like a map, agree on certain meanings expressed by certain signs. This system of signs is rational and undebatable.[9]

Bertin makes a distinction between the content of a map, which he calls the information, and the container, the graphic system. The number of components used in the graphic system depends on the number of elements the information constitutes. As an example, Bertin presents a map of the Iberian Peninsula in which three values of population density are used.[10] The information in the map consists of three elements and therefore the graphic system should also show three values. Bertin distinguishes three types of graphic marks—point, line and area—that can vary in terms of position, size, value, texture, color, orientation and shape. He calls these the "visual variables."[11] In the example of the map of Spain and Portugal, a point is used as graphic mark to represent the population density and size is used to express different values in population density. Small dots are used in areas with a low population density and big dots in denser parts. Bertin explores the manipulation of the visual variables with the aim to define rules of construction and legibility in order to improve the efficiency of maps and other visualizations. His methods are aimed at reduction and simplification, based on a rigorous analysis of the components of the information.[12] Bertin wants to create clear and efficient messages.

A Bertinian Reading of Google Maps

In this section I will take the ideas of Bertin to evaluate Google Maps. In doing so I not only want to better understand and unearth flaws in Google Maps, but also find out if Bertin's theories, developed in a time of paper cartography, are applicable to cartographic apps as well.

Using Bertin's mathematically rational considerations, it is not difficult to find shortcomings in Google Maps. For one, the geospatial application does not provide a legend of its map. It is not possible for a user to have a clear understanding of the meaning of the map's colors and signs. Why is it, for instance, that the Haagse Bos, a landscape park in The Hague, is shown in gray in Google Maps while surrounding parks like Koekamp, Malieveld, Oostduin, Landgoed Clingendael, Oosterbeek and Haagse Hout are shown in green?[13] The Amsterdamse Bos, a landscape park in Amsterdam, on the other hand, is green on Google Maps.[14] The Kralingse Bos, a large landscape park in Rotterdam, is as gray as its counterpart in The Hague, but has some green patches in it.[15] Zooming in, and comparing the map with the satellite view of the same area in Google Earth, it appears that these green patches are sport fields. Strangely, in the map of the Kralingse Bos the football pitches appear green, while the athletics track with a similarly sized grassy center field for javelin, shot put and other nontrack events, is gray.

Although Google Maps fails to provide a legend, it does offer users clues by providing more detailed information and alternative representations. The geospatial application is a dynamic platform that lets users zoom in and out and change the map modus to the Google Earth setting, which contains satellite imagery, as well as offering Google Street View, photographic panoramas of stitched images recorded from positions along streets.

6   Eckert, "Is Innovation in Cartography a Mere Illusion?"
7   Eckert and Jégou, "Quel planisphère de références pour Google Maps."
8   Bertin, *Semiology of Graphics: Diagrams Networks Maps*, 2.
9   Ibid., 3.
10   Ibid., 20.
11   Ibid., 7.
12   Ibid., 171.
13   Google Maps, "Haagse Bos," accessed 3 May 2019, https://www.google.com/maps/@52.0942603,4.3365666,15z.
14   Google Maps, "Amsterdamse Bos," accessed 3 May 2019, https://www.google.com/maps/@52.3185022,4.8331301,15z.
15   Google Maps, "Kralingse Bos," accessed 3 May 2019, https://www.google.com/maps/@51.938218,4.5144516,14.75z.

The Legend of Google Maps: Zoom Levels

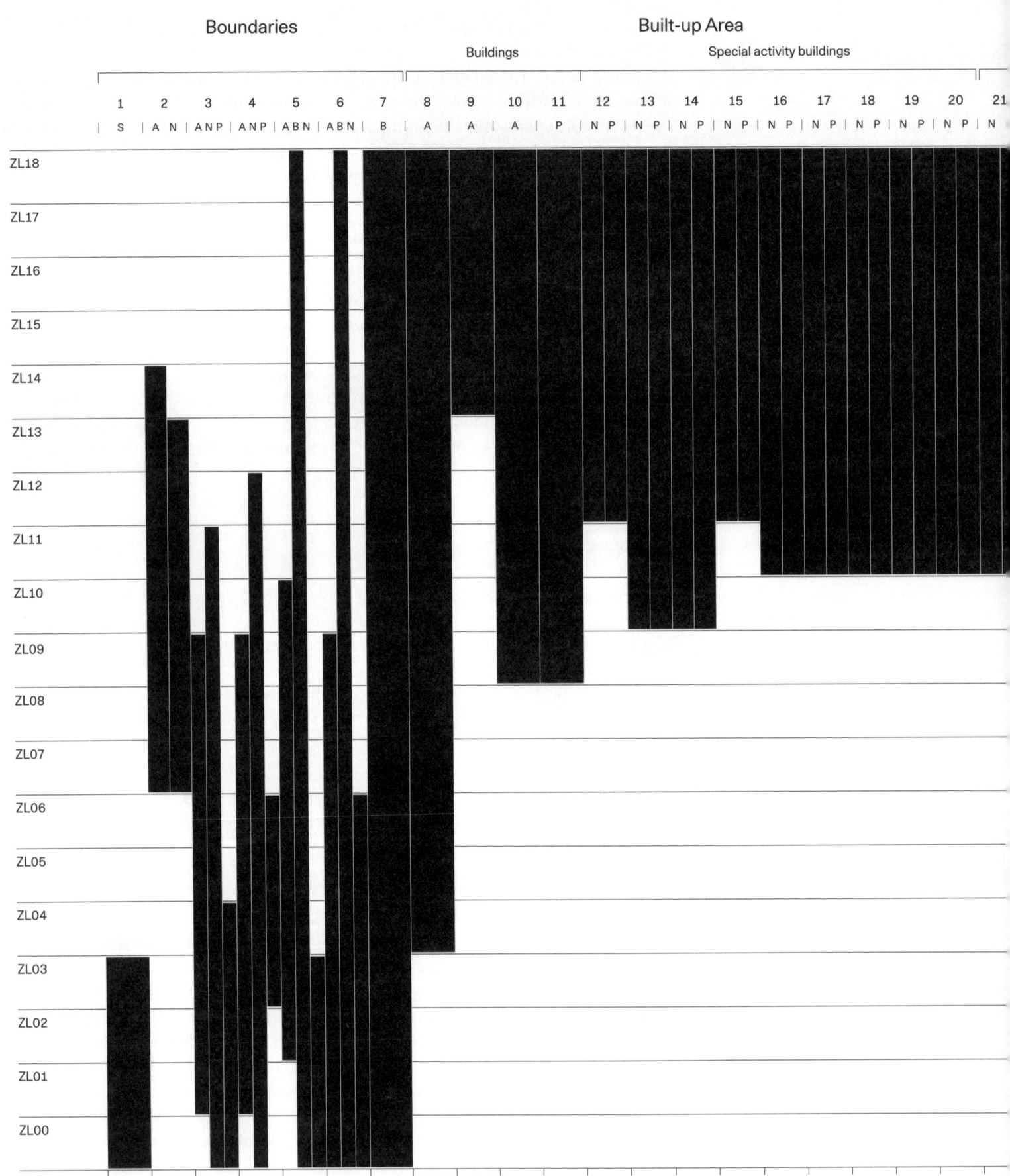

This chart shows which information Google Maps contains on each of the nine-teen zoom levels. Google Maps contains significantly more information on the closer zooms than on the more remote zooms. The cartographic typology of Google Maps changes per zoom level, from city map to road map to topographic map. This table shows that this is a gradual process.

Traffic Network   Hydrography   Topography   Google

Transports   Roads

22	23	24	25	26	27	28	29	30	31	32	33	34	35	36	37	38	39	40	41	42	43	44
P	N P	S	N D S	N D S	N D S	S	A N	A N	A N	A N	N S	A	A N	A	A	A N	A N P	N P	N P	N P	N P	P

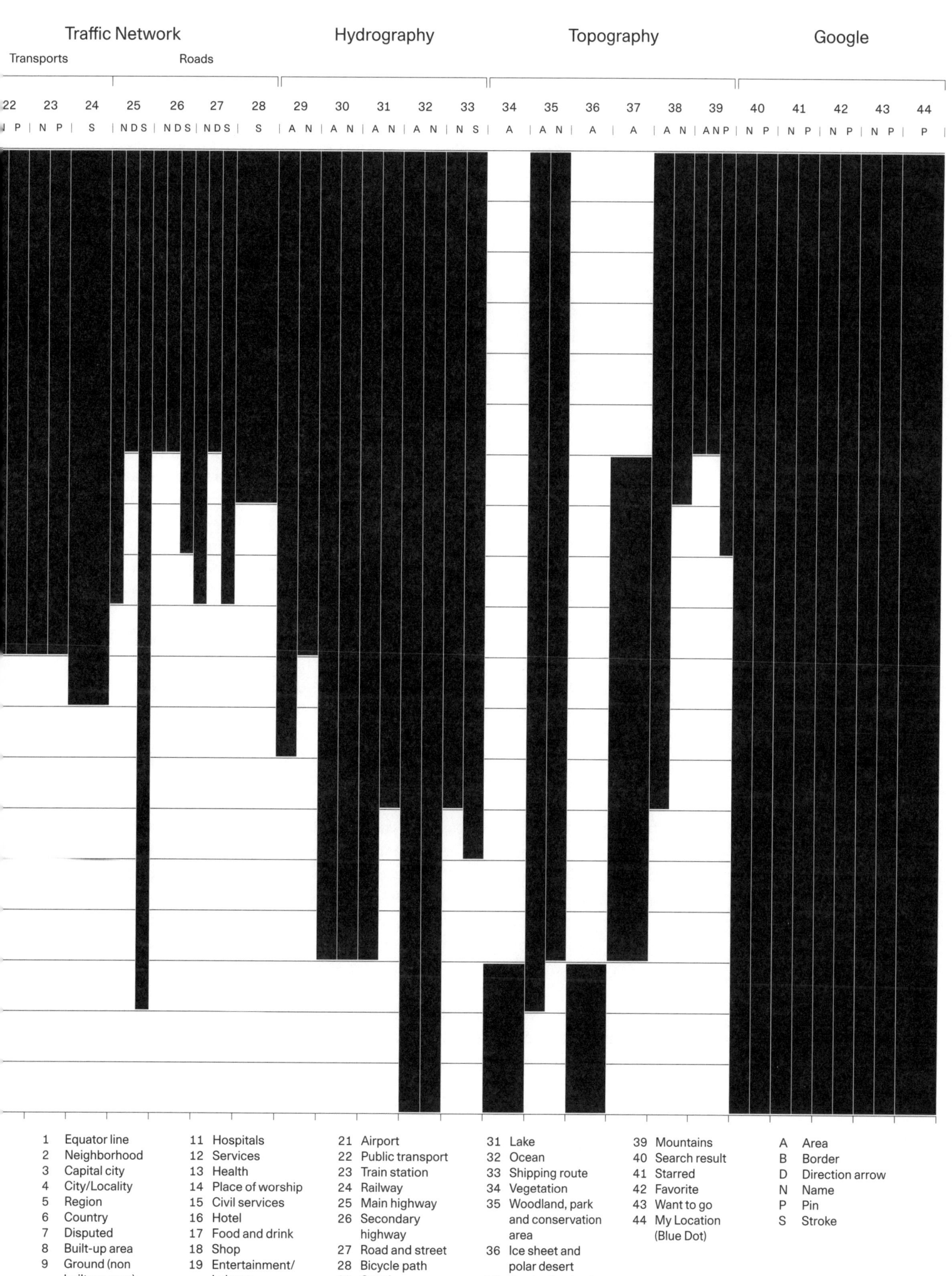

1	Equator line	11	Hospitals	21	Airport	31	Lake	39	Mountains	A	Area
2	Neighborhood	12	Services	22	Public transport	32	Ocean	40	Search result	B	Border
3	Capital city	13	Health	23	Train station	33	Shipping route	41	Starred	D	Direction arrow
4	City/Locality	14	Place of worship	24	Railway	34	Vegetation	42	Favorite	N	Name
5	Region	15	Civil services	25	Main highway	35	Woodland, park	43	Want to go	P	Pin
6	Country	16	Hotel	26	Secondary		and conservation	44	My Location	S	Stroke
7	Disputed	17	Food and drink		highway		area		(Blue Dot)		
8	Built-up area	18	Shop	27	Road and street	36	Ice sheet and				
9	Ground (non	19	Entertainment/	28	Bicycle path		polar desert				
	built-up area)		Leisure	29	Canal	37	Production land				
10	Areas of interest	20	Outdoor	30	River	38	Beach				

67

In my view, Google does not offer a legend because its focus is on the display of addresses rather than on showing the landscape. The Maps app is a free software tool that up to now does not bring in revenue for Google. The company does not disclose how it intends to monetize the app, but in a 2017 interview Google CEO Sundar Pichai suggested this might be achieved through advertising.[16] Google developed a method of raising revenue by selling advertisements in its search engine. An algorithm analyzes every search and determines which advertisements will be shown as "sponsored links" at the top of the search results.[17] Just as information about what we search is valuable to advertisers, so can be the information about where we search, or what we search in certain locations. The map of Google contains the location information of companies, shops and restaurants that in the future might be willing to pay for users to find their whereabouts easily. So it is in Google's interest to make these locations clearly visible on the map and to only minimally indicate other data, such as information about different types of buildings or green areas. This hierarchy of information becomes apparent in how the map is structured. Google Maps consists of two layers. The background layer, a base map containing topographic information, is a geographic support for a toponymic layer of names of places, companies, shops, restaurants, stations, streets and cities. The hierarchy between the layers is apparent from the visual strategies of the map design. The topographic layer is rendered in pale versions of colors often used in maps: gray for the built environment, green for vegetation and blue for water. Google Maps looks familiar but vague: a mere memory of a map. In contrast, the foreground information, that is the names and pictograms of the information Google might intend to monetize, is shown in bright colors. These fundamentally diverse data sets show up differently at distinct zoom levels.

Here we touch, in my view, on a second difference between Bertin's thinking about maps and Google's geospatial application. Whereas to Bertin a map is an unambiguous stable rendering of the world, Google maps is a dynamic, mutating representation. The plural in the name Google Maps is slightly confusing, as the application contains but a single map that can be scrolled endlessly in either direction. It can also be zoomed in or out and for each of the nineteen zoom levels, a different map is rendered. The map is based on a database with certain elements showing up at particular zoom levels only. Each of the zooms contains different information and is a unique map. Over the course of the zoom levels the digital map changes in typology: from street map to road map to topographic map. To me, this ability to mutate is an aspect never seen before in maps on paper. In that sense the plural in the name Google Maps is correct after all.

A result of showing different kinds of information at different scales is that Google Maps at certain zooms seems to contain hardly any information, while at others too much. This becomes apparent when looking at a shopping street on a small scale—a map in a small scale shows objects closer than a map in a large scale—the map looks too full. Dozens of names and symbols cover the map. Texts are rendered in different directions, horizontal for places like shops, restaurants and institutions, in angles following roads for street names. Furthermore, the texts are set in a small type size to avoid overlapping. The pale colors used for the topographic backdrop do not provide structure to the image.

By contrast, looking at the same location zoomed out, the map feels too empty. With each step up names of buildings, streets, cities, countries appear and disappear. At a certain zoom level no new information appears. What remains is the topographic backdrop that feels quite naked when its pale hues are not covered by texts. Just like the closeup zoom, the map at this larger scale level does not offer overview. It is suspiciously empty.

The necessity for having different zoom levels comes from the poor resolution a screen offers in comparison with printed matter. That was especially the case in the early days of Google Maps when the size and resolution of displays—particularly those of smartphones—was much lower than today's high-resolution screens. Still, even today's devices are lagging behind in resolution compared with ink on paper. On my laptop's 13.3 inch retina display, a screen with an especially high resolution, a map will show the same amount of detail as a printed image of 21.9 × 13.5 cm. A map in a typical atlas, for instance the Dutch *Grote Bosatlas*, is five times bigger in surface area. A large tablet computer has an effective surface area that is seven times smaller than a *Bosatlas* map. And a standard smartphone shows nearly thirty times less information than the printed atlas. The digital map is not able to give its user the same overview as a printed map. Instead, it offers dispersed content through a sequence of zooms.

In my view, a further point of critique of Google Maps from a Bertinian perspective would be its editorial fuzziness. There is no clear relation between the crop of the map, the information displayed and the visual variables employed.

Google Maps is made using large quantities of data that are being updated constantly. To speed up the production of the map and its distribution via the Internet, the geographic plan is divided in smaller segments called map tiles. Each of these tiles in Google Maps measures 256 × 256 pixels. For each zoom level a different map is produced. At the outermost zoom level, level zero, the entire world is rendered in a single map tile. Every zoom level up, one tile is replaced by four tiles. This means that at closer zoom levels, Google Maps contains millions of map tiles. As a consequence and for understandable reasons, the editorial and design processes that Bertin sees as essential in the making of maps have been replaced by a programmed set of instructions.

Even if it would be feasible to carefully craft the map through programming, it would not be possible to anticipate the crop that a Google Maps user will choose. As a result, Google Maps shows a map view that is unbalanced in terms of the quantity, quality and density of information. But Google Maps should not be regarded as a static image; it is a digital platform that offers users the zoom and crop of their choice, and, if they wish, finds them additional information such as directions, opening hours and phone numbers. What Google Maps lacks in informational structure it compensates, I would argue, in offering interactivity and providing additional information.

Predigital Practice

Reading Bertin with a computer running Google Maps at hand, switching back and forth between the page and the screen to read and test, it becomes apparent

16 Orerskovic, "Sundar Pichai Just Hinted at How Google Will Make Money from Maps, and it Sounds Like Lots of Ads."
17 Brotton, *A History of the World in Twelve Maps*, 428.

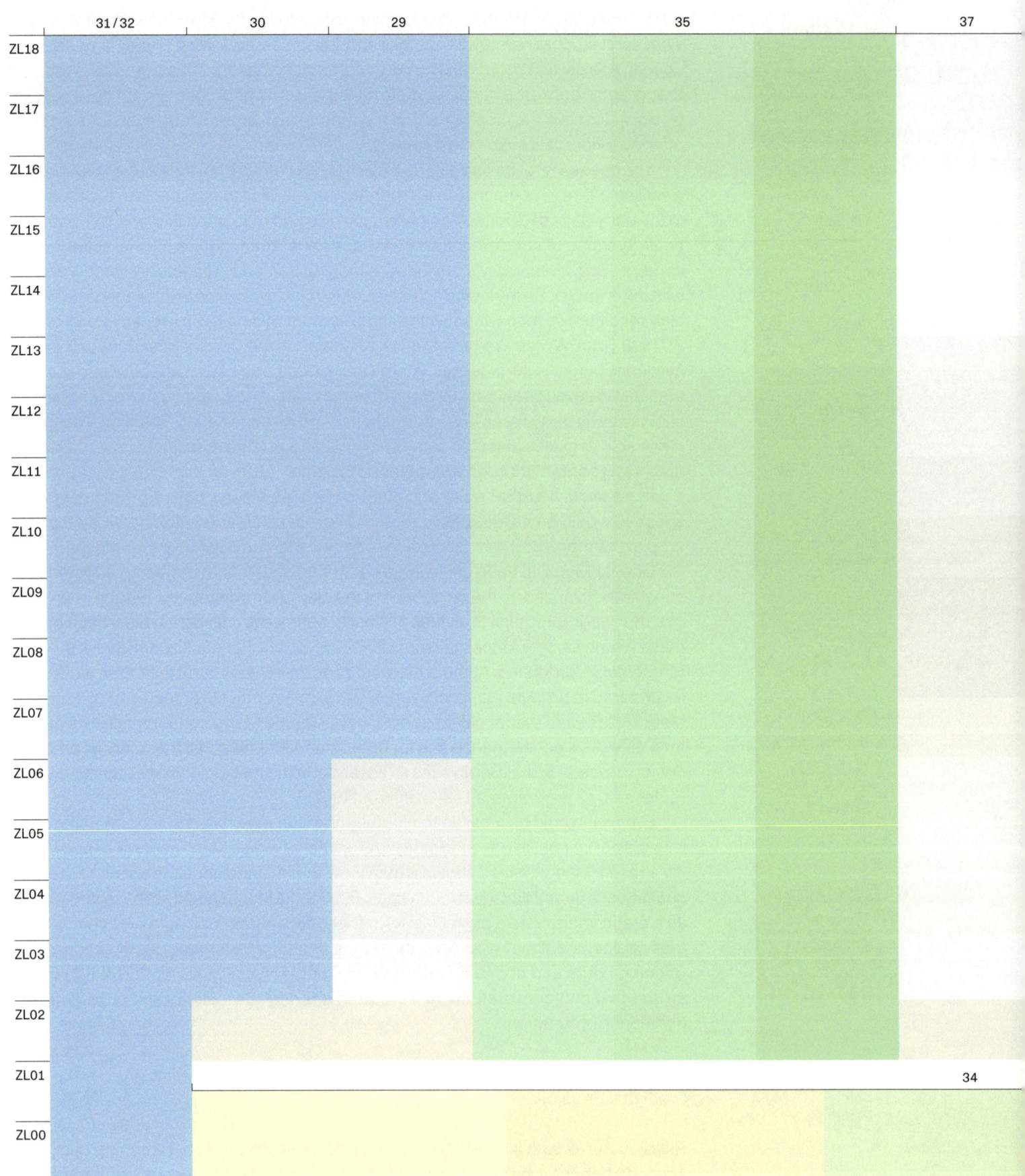

This chart, which is a combination of 3.1 and 3.2, shows the legend of the base map of Google Maps: the colors of the topography, hydrography and built-up areas, organized vertically by zoom level. Note that the colors are pale and soft, this in contrast to the information on specific places and search results as shown in 3.4.

38 8 9 10 11 36 39

1 Equator line
2 Neighborhood
3 Capital city
4 City/Locality
5 Region
6 Country
7 Disputed
8 Built-up area
9 Ground (non built-up area)
10 Areas of interest

11 **Hospitals**
12 Services
13 Health
14 Place of worship
15 Civil services
16 Hotel
17 Food and drink
18 Shop
19 Entertainment/ Leisure
20 Outdoor

21 Airport
22 Public transport
23 Train station
24 Railway
25 Main highway
26 Secondary highway
27 Road and street
28 Bicycle path
29 Canal
30 River

31 Lake
32 Ocean
33 Shipping route
34 Vegetation
35 Woodland, park and conservation area
36 Ice sheet and polar desert
37 Production land
38 Beach

39 Mountains
40 Search result
41 Starred
42 Favorite
43 Want to go
44 My Location (Blue Dot)

71

that Bertin's practice is a predigital one grounded in practical experimentation. In his *Semiology of Graphics* many references are made to predigital tools and skills and materials like tracing paper, typewriter and handwriting (all page 308), "sureness of hand of the professional draftsman" (page 311) and pencil and ink (both page 312). This realization nuances the universality of some of Bertin's claims beyond the predigital period.

Look, for instance, at Bertin's thoughts on color. Bertin points out that next to not being able to reach the color-blind, the disadvantages of the use of color are reprographic problems like additional production time and costs of multicolor documents over monochrome production.[18] This argument might have been valid in 1967 when Bertin wrote *Semiology of Graphics*, but nowadays it is hardly the case in the production of printed matter and completely irrelevant in production for screen use.

Bertin also dwells on the difficulties of creating a color range of equal value. Certain colors, like yellow, are much lighter than blue and red, for instance.[19] Rather than colors, Bertin prefers the use of monochromatic patterns. Through difference in texture—Bertin uses the French word *grain* to indicate the fineness of the pattern—a variety of shades can be realized. This is a common technique in printing, but patterns do not display well on screen because of the coarse resolution of monitors. It therefore makes sense that as a digital application designed to be looked at on screen, Google Maps uses color as a visual variable rather than a monochromatic pattern.

Another issue addressed by Bertin that seems to be a bigger challenge in a static print than in a dynamic screen application is map generalization. This is an abstraction method to create a smaller-scale map from a larger-scale map or from map data with a higher level of detail. This can be done manually by a cartographer, or through an algorithm. In *Semiology of Graphics* Bertin spends quite some pages addressing the matter. As a case study he compares nine maps of the Dombes area between Lyon and Bourg-en-Bresse.[20] The region contains more than 1,000 small ponds and lakes that show up as a pattern of small dots on the maps. Bertin notes that the generalization of a cluster of marks, like the pattern of lakes in the Dombes, is the most complex in its kind. The nine maps show a great difference in the number of lakes and their shapes and sizes. The 1:1,000,000 map from the *Times Atlas of the World* contains twenty lakes. The one from *Atlas de France—Geomorphology* shows more than 100. Looking at the same region in Google Maps, the number of lakes displayed depends on the scale of the map. At a certain zoom level only eleven lakes are visible. One zoom level closer the map shows more than 200 patches of blue. Generalization seems less of an urgency in the context of geospatial application when with a touch of the mouse, or pinch on the screen, more detailed information can be accessed. Overview in the case of the digital map is not a singular high-resolution image, but scrolling through a sequence of disparate views.

In summary, it can be said that Google Maps has certain limitations. These are partly related to choices made in favor of technology over cartography. Furthermore, it is clear that Google has developed its map app with a view to future earnings. It puts emphasis on showing (commercial) addresses over topography. This becomes apparent in the color choices made in the design and probably also explains why Google Maps withholds certain basic cartographic information such as the legend. On the other hand, Google Maps offers certain functional improvements over paper maps.

By systematically studying the communicative aspects of graphics, Bertin has vastly deepened the understanding of map design. As his studies are rooted in the pragmatics of a predigital practice, the notions and methods Bertin developed mainly apply to printed matter and feel in some instances inadequate to evaluate a cartographic platform like Google Maps.

The work of Bertin is part of a larger set of practices aimed at improving map effectiveness. This approach to mapmaking has been called cognitive cartography[21] and representational cartography.[22] The premise of this method is that the world can be objectively known and truthfully mapped. American geographer and cartographer Arthur Robinson (1915–2004) is another noteworthy representative of this approach to mapmaking. Using ideas of cognitive psychology Robinson aimed to improve map designs by carefully controlled scientific experiments such as how to best represent location, distance and direction.

A Critical Reading of Google Maps

In this section I will look at a critique of cartography that emerged around the time that digital mapping tools became available. In response to some of the notions raised by this critical approach, I will propose a strategy to designing maps developed in my practice.

At the end of the 1980s, ideas from poststructuralist thinking, social constructionism and actor-network theory resulted in a shift in the thinking about maps. These new practices have been described as more-than-representational cartography[23] and critical cartography.[24] British geographer, cartographer and map historian Brian Harley (1932–1991) is one of its main representatives. Influenced by the work of Michel Foucault and Jacques Derrida, Harley regarded a map to be a social construction.[25] Building on Foucault's notion about the omnipresence of power in all knowledge, albeit invisible or implied, Harley states that maps are the products of power but also produce power themselves.[26] Derrida's 1967 book *Of Grammatology*, original title *De la grammatologie*, a foundational text of deconstructive criticism, instigated Harley to look at the textuality of maps, particularly their rhetorical dimension. To Harley, mapping is not a neutral, objective pursuit, but one laden with power. It deals with creating knowledge and not with revealing it. In the 1990s, Harley's work initiated the interdisciplinary field of critical cartography. Its objective is to study the full scope of the map, both as the result of a process and as a communicative object.

Geographers Jeremy W. Crampton of Georgia State University, Atlanta, and John B. Krygier of Ohio Wesleyan University, Delaware, connect the development of a critical reading of cartography to the digital mapmaking practices that emerged. They describe two circumstances that resulted in "cartography slipping from the control of the powerful elites"—such as the great map houses of the West, the state, and to a lesser extent the academic world—that have dominated it for several hundred years.[27] Digital technology, new mapmaking software, open-source

18 Bertin, *Semiology of Graphics: Diagrams Networks Maps*, 91.
19 Ibid., 89.
20 Ibid., 307.
21 Caquard, "A Post-Representational Perspective on Cognitive Cartography."
22 Kitchin, "The Transformation of Cartographic Thought."
23 Ibid.
24 Crampton and Krygier, "An Introduction to Critical Cartography," 12.
25 Harley, "Deconstructing the Map," 57.
26 Ibid.
27 Crampton and Krygier, "An Introduction to Critical Cartography," 12.

This chart, which is a combination of 3.1 and 3.2, shows the legend of the super-imposed information in Google Maps: specific places and search results. Note that the colors of this map are much brighter and bolder than that of the base map 3.3.

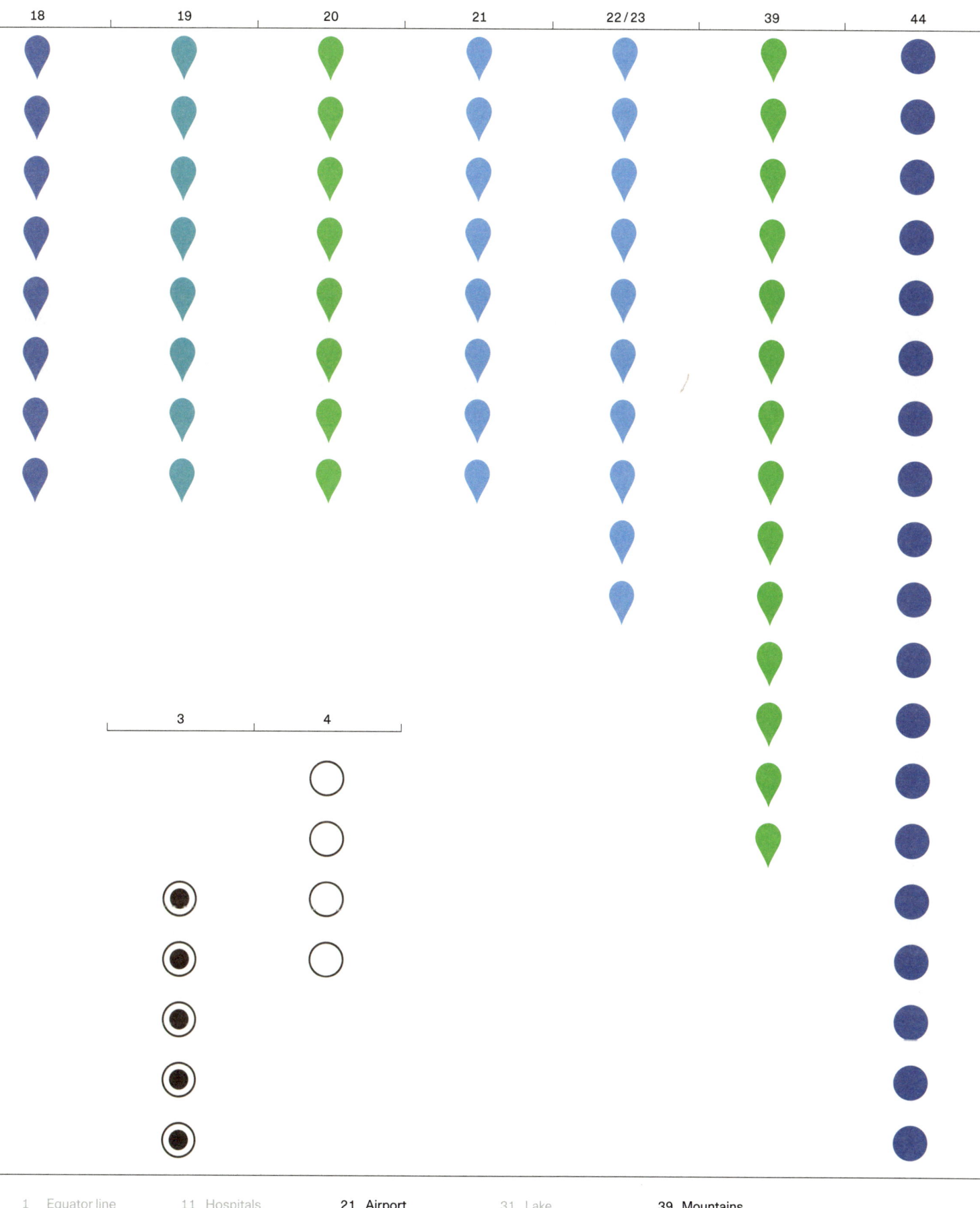

18	19	20	21	22 / 23	39	44

3	4

1	Equator line	11	Hospitals	21	Airport	31	Lake	39	Mountains
2	Neighborhood	12	Services	22	Public transport	32	Ocean	40	Search result
3	Capital city	13	Health	23	Train station	33	Shipping route	41	Starred
4	City/Locality	14	Place of worship	24	Railway	34	Vegetation	42	Favorite
5	Region	15	Civil services	25	Main highway	35	Woodland, park	43	Want to go
6	Country	16	Hotel	26	Secondary		and conservation	44	My Location
7	Disputed	17	Food and drink		highway		area		(Blue Dot)
8	Built-up area	18	Shop	27	Road and street	36	Ice sheet and		
9	Ground (non	19	Entertainment/	28	Bicycle path		polar desert		
	built-up area)		Leisure	29	Canal	37	Production land		
10	Areas of interest	20	Outdoor	30	River	38	Beach		

collaborative tools, mobile mapping applications and geotagging opened up the field of cartography and made it possible for new mapmaking practices to emerge. The second circumstance was the forming of a critique that highlighted the politics of mapping, with the aforementioned Harley as one of its key protagonists. According to Crampton and Krygier, the critical approach to cartography "undisciplined" the field, freed it from the confines of the academic and opened it up to the people. Until the digital age, maps were produced by those in power, such as the State. In my view however, the new tools created new power centers in companies such as Google that owned and controlled the new technologies. According to Crampton and Krygier, digital technology created new cartographic tools that are easy to access. Technology thus empowered new players to enter the field who, with no prior knowledge of cartography, started mapping different subjects, in novel ways, occasionally resulting in original forms.

Looking at Google Maps from a critical cartography point of view, various power mechanisms and manipulations become apparent, some visible, others hidden. An example of a manipulation that is visible is the map projection, the method of how a spatial object such as Earth is translated into a two-dimensional map. When Google Maps was launched in 2005 it used the Mercator projection, named after Flemish geographer and cartographer Gerardus Mercator (1512–1594), who invented it in 1569. In the Mercator projection all meridians run parallel, creating a rectangular map of the spherical Earth, which results in the distortion of the size of objects on latitudes further away from the equator. Land masses near the poles, such as Greenland, appear much larger than land masses near the equator, such as Central Africa. Projection methods that cause distortions, such as that of Mercator, have been accused of cartographic imperialism, as they depict the richer, powerful countries bigger than less-wealthy countries.[28] In Google Maps the distortion caused by the Mercator projection is mainly visible in the most zoomed-out views. In August 2018, Google Maps dropped the Mercator projection for the most zoomed-out views and instead shows a three dimensional globe.

Depending on the part of the world from where you access Google Maps, the cartographic app will show different, or adapted, information. This is, for instance, the case with contested regions. The Crimean peninsula on Google Maps, when accessed in the Netherlands, is separated from Ukraine by a black, dotted line indicating it as a disputed border.[29] The same graphical representation is used for the borders of the Gaza Strip, the West Bank and between North and South Cyprus. Looking at the Crimean-Ukrainian boundary on a Russian Google Maps, however, the line is a solid, black country border indicating that Crimea is part of Russia.[30]

In a 2016 blog post, Justin O'Beirne, who writes about mapping apps, compares Google Maps from 2010 and 2016.[31] O'Beirne notices that the more recent map contains far fewer city labels and names of cities and more roads in comparison with the same map in 2010. In the period between the releases of the two maps, the usage of Google Maps on mobile devices surpassed that on desktop computers.[32] O'Beirne therefore speculates that the reduction of city names displayed in Google Maps is an optimization for reading the map on mobile devices, the new dominant technology, and that the new roads were added to make the map look less empty. A map with fewer labels can be read faster. The reasons to do so might be understandable, but taking out one kind of information and replacing it with another is manipulation.

From a critical cartography point of view, the examples above—the Mercator projection, the different display of disputed areas depending on the location of the user, and the inclusion of less information to optimize a map for a new dominant technology—are manipulations of a map's content that are visible signs of how a map produces power. There are also imperceptible aspects that highlight that Google Maps is also a product of power. In October 2004, Google took over Keyhole, a geospatial data visualization company. Keyhole was co-owned by the US foreign intelligence service Central Intelligence Agency (CIA) and its major clients were the US military and intelligence agencies. Bearing this in mind, there is a brutal honesty in the name Keyhole that refers to a device to peek through and spy on others. Keyhole's software Earth Viewer formed the basis for Google Earth. Other aspects of Earth Viewer were integrated into Google Maps. In 2003 American news television channel CNN used Earth Viewer software extensively in its reportages of the invasion of Iraq, showing images that simulated flying over Baghdad and dropping down to street level at bombing targets.[33] This was a new form of war reporting financed by the military. Two years later, with the launch of Google Earth, this new form of cartographic imagination would be accessible to everyone.

Other technologies used by Google in its cartographic software were originally developed for military use as well. The satellite imagery used in Google Earth and the Global Positioning System (GPS) used in the Google Maps mobile app were initially developed by the military for surveillance and navigation purposes. There will be more about this subject in the chapter on the Strava Global Heatmap.

Another aspect of how Google Maps produces power is that it is a technology embedded in an economic model based on surveillance. Google maps the users of Google Maps. The technology company records data of the users of its free services like its cartographic app. What users search, on what device, when, where and who they are: it is information that the technology company stores and analyzes. Google monetizes the acquired data through targeted advertisements. This economic model has been called surveillance capitalism by Shoshana Zuboff, a professor emirata at Harvard Business School.[34] In that sense the "maps" in Google Maps not only refers to a noun plural, the variety of maps that make up the cartographic app, but also to "maps" as a verb, as in Google that actively records the actions of its users.

It is not only Google that monitors the users of its Maps app. In 2014 documents leaked by former National Security Agency contractor Edward Snowden reveal that intelligence agencies NSA, the United States National Security Agency and GCHQ, the United Kingdom Government Communications Headquarters, intercept Google Maps searches made on smartphones and store these with the location information of where the query is made. This was so successful that a 2008 document points out that "it effectively means that anyone using Google Maps on a smartphone is working in support of a GCHQ system."[35]

Ambiguous Strategies

The aforementioned Eckert doubts if it is possible to produce truly new maps. He may be right, if the map is regarded as functional object alone. From that point of view a map's efficiency can be improved through cognitive scientific experiments.

28 Cartography and imperialism have been linked for centuries. Maps have been instrumental in the overseas expansion of European powers as source of geographic knowledge and as tools for the planning of human settlements.

29 Google Maps, "Crimean peninsula," accessed 12 May 2019, https://www.google.nl/maps/place/Crimean+Peninsula/@45.6398034, 33.246309,7.63z/data=!4m5!3m4!1s0x40 eac2a37171b3f7:0x2a6f09e02affbaeb!8m2 !3d45.3453029!4d34.4997274.

30 Usborne, "Disputed Territories: Where Google Maps Draws the Line."

31 O'Beirne, "What Happened to Google Maps?"

32 Siegler, "Google Maps For Mobile Crosses 200 Million Installs; In June It Will Surpass Desktop Usage."

33 Maney, "Tiny Tech Company Awes Viewers."

34 Zuboff, "Google as a Fortune Teller: The Secrets of Surveillance Capitalism."

35 Ball, "Angry Birds and 'Leaky' Phone Apps Targeted by NSA and GCHQ for User Data."

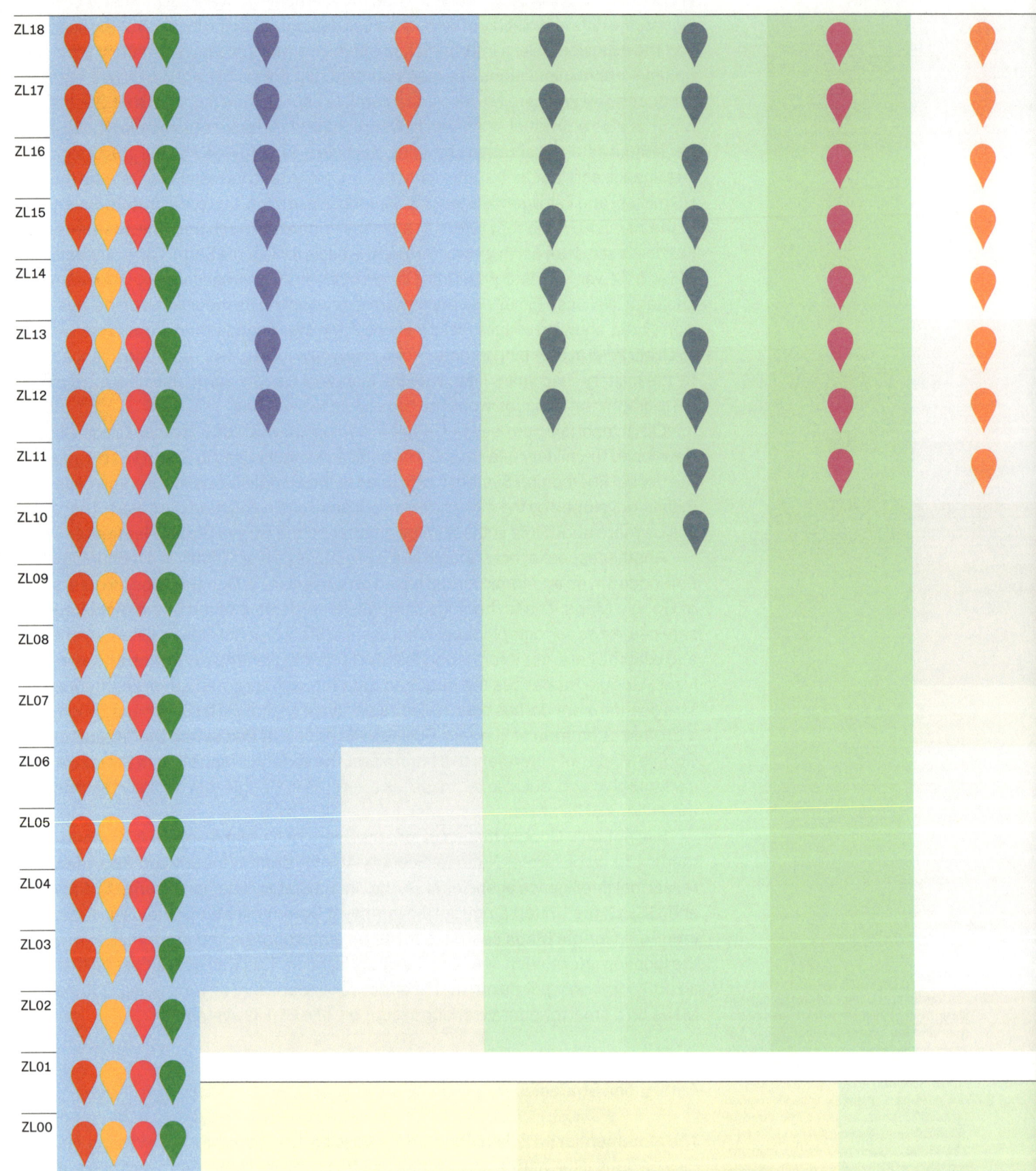

This chart, which is a combination of 3.3 and 3.4, shows the bright- and bold-colored information of specific places and search results superimposed on the pale, soft-colored base map.

The methods to achieve this by the earlier-mentioned Bertin are impressively thorough and probably unsurpassable in improving this aspect of the paper map. However, if the focus were to shift from the communicative aspect of maps to an approach that incorporates critical insights, then, perhaps, it would be possible to create innovative maps. The aim would be to show users that maps are the product of a process that is not a neutral, objective pursuit, but one laden with power.

What strategies can be used by the map designer to show that a map is not a neutral object? American visual theorist and culture critic Johanna Drucker suggests that this could be achieved through highlighting the ambiguous nature of the map. Drucker regards information visualizations such as maps as "intellectual Trojan horses" of the empirical sciences that suggest they are observing independently, but are in fact interpretations masquerading as representations.[36] According to Drucker we need to accept the fact that the nature of data is fundamentally constructed and acknowledge that phenomena such as nations, genders, populations and historical periods are not self-evident, stable entities. To rethink our approach to visualization and to the assumptions that underpin it, Drucker proposes that ambiguity and uncertainty ought to be incorporated in the design of information, either by representing ambivalence or by using ambiguity as the basis on which a representation is made.[37] More concretely, Drucker calls for a map with a more nuanced legend or for a nonstandard map that shows its constructedness.

In my design practice I have developed a number of strategies to show the ambiguous nature of the map. Some focus on introducing nuance in the legend; others use the means of production to highlight the cartographic construct.

In a recent project my studio worked on for Inrap, the French National Institute for Preventive Archaeological Research, visual strategies were developed for a series of maps that underline the uncertainty of the displayed data.[38] In two maps that show how *Homo sapiens* spread over the world, line pattern transitions are placed between surfaces that display occupation during certain time periods. The same visual transitions are used in timelines placed under the maps. The line patterns create soft outlines around the shapes depicting territories or historical periods. The soft edges are a visual metaphor for doubt that highlights that the information on the map is but a hypothesis. The blurry boundaries also indicate that the informations on the maps and in the timelines are static renderings of dynamic processes.

Primary and secondary colors were avoided in the choice of colors for the archaeological maps. My problem with yellows, reds and blues is not the difference of their retinal weight (as Bertin would have it), but rather that primary colors constitute a complete model, in which there is no room for a fourth or fifth primary color. Displaying cartographic information in these three colors therefore suggests completeness, whereas it was our aim to do precisely the opposite, namely to cast doubt on whether the information was complete and unequivocal.

In another project, *Atlas of the Copenhagens*, ambiguity was introduced by using a different map typology for each chapter rather than using a singular cartographic representation throughout the book.[39] One of the qualities I appreciate about Google Maps is that it mutates from zoom level to zoom level. Zooming out, it starts as street map, becomes road map and transforms into topographic map. We aimed to achieve a similar mutation in the Copenhagen atlas: each of the six chapters of the atlas uses a different type of map, from descriptive geographic maps, to data maps, to isometric bird's-eye view drawings, to photographic assemblages. By showing a range of maps that are very different typologically and varied in their visual presence, the territorial and conceptual limits of a city are questioned. This questioning is highlighted by the plural "Copenhagens" in the book's title. As well as being an atlas that scans a city in different ways, the book is also a catalogue of various representation types.

Atlas of the Copenhagens is printed in two fluorescent colors, a very strong blue and a black. The inks emphasize the materiality of the printed object. Fluorescent inks are less transparent than regular inks and lie, literally, more "on top of" the paper. This gives them a stronger physical presence. This means that a map printed in fluorescent inks has a heightened materiality in comparison to a map printed in regular inks, where paper and inks are physically more one. The effect of the inks emphasizes that a map is both a representation of a physical reality and a material object itself.

Letting maps bleed off the page is a further strategy that uses the means of production to emphasize the manipulation in the process of making maps. A map is a cut-out of reality. This is highlighted by emphasizing the physical nature of the map. The cut of the paper in the production process becomes the crop of the map.

In my work I emphasize and challenge cartographic manipulations by using the means of production to highlight these very manipulations. Making maps is a process of controlling, distorting and altering data. Manipulation takes place in projecting a spatial reality on a two-dimensional plane, in choosing an orientation of a map, in making it a fixed rendering of a situation in flux, in resizing an original site to a smaller representation, in showing a cut-out version of reality, in filtering elements by selecting them or generalizing them, and in adding invisible elements like borders and height lines. A similar transformation occurs in the use of the map. The user is manipulated through the result of the process of reproduction and multiplication. In the choice of medium, the physical qualities, the glow and brightness of a screen or the tactility and translucency of paper. In how a map is placed on a sheet or screen, bleeding out, with a margin, or in a composition with other elements. In the use of typography. In the choice of reproduction technique, the physical qualities of inks on paper, the fact that certain RGB-recipes produce colors that are more intense on the screen. In the effect of grids, line weights, and other characteristics of reproduction.

In November 2017 I presented my work and ideas about an ambiguous cartography as described above at the conference "Graphic Design and Research in the Social Sciences. Jacques Bertin and the Graphic Laboratory. EHESS 1954–2000" at the École des Hautes Études en Sciences Sociales (EHESS) in Paris.[40] The conference was organized on the occasion of the fiftieth anniversary of Bertin's book *Semiology of Graphics*. My presentation did not get much response from a large part of the audience. It was as if I was addressing an issue that was not a problem to this audience. A year later, I received a more positive response when I presented my work at the "Mappings as Joint Spatial Display" conference in Berlin.[41] The occasion was very different. Rather than celebrating a theoretical legacy, the symposium in Berlin was aimed at furthering a methodological discourse in mapmaking and

36 Drucker, *Graphesis: Visual Forms of Knowledge Production*, 125.
37 Ibid., 127.
38 Demoule, Garcia and Schnapp, *Une histoire des civilisations: Comment l'archéologie bouleverse nos connaissances*, 34–45.
39 Simpson, Gimmel, Lonka, Jay and Grootens, *Atlas of the Copenhagens*.
40 "Design graphique et recherches en sciences sociales: Jacques Bertin et le Laboratoire de Graphique. EHESS 1954–2000."
41 "Mappings as Joint Spatial Display."

French cartographer and theorist Jacques Bertin distinguishes three types of graphic marks—point, line and area—that can vary in terms of position, size, value, texture, color, orientation and shape. He calls these the "visual variables." The overview on the left-hand page shows the variety of visual variables. Bertin included similar overviews in his publications. The right-hand page shows the visual variables of Google Maps, in which only a limited part of the width of the variables is used. Some of Bertin's variables, such as texture, are focused on printed matter and therefore not suitable for screen applications such as Google Maps. Other variables, such as color, are used only to a limited extent. In Google Maps, the background colors of the base map are pale, to emphasize the brightly colored location information placed on top of it.

in developing instruments for spatial research. The Berlin conference brought together scholars and students from a variety of fields: sociology of space, architecture, urban studies and geography. I felt that at times my work acted as a bridge in this interdisciplinary debate. The critical concerns of my research resonated with the social scientists, whereas the typologies and tools that I proposed connected with the architects and urban designers. Looking back at the two conferences, I feel that the critical approach to maps may not be a next phase, but rather presents a bifurcation, a parallel trajectory in the thinking about cartography.

A Post-Representational Reading of Google Maps

In this section I will look at how recent cartographic thinking sees maps as processes rather than as products. In this line of thinking, cartography is redefined as a broad set of practices that incorporate both the production and application, the producer and the user, of the map. With this as frame of reference I will investigate the Blue Dot in the Google Maps app, how it functions, how it was introduced and the technological contexts surrounding it. I will also speculate on alternative designs.

After representational cartography, which emphasized the map as object of communication, and more-than-representational cartography, which saw the map as a product of power but also as a producer of power, the early 2000s saw a third shift in the thinking about maps called post-representational cartography. In post-representational cartography the fundamental status of the map is questioned. Maps are seen as never fully formed, their work never complete.

In the previous chapter I introduced the term *Blind Map* for maps (and other visualizations) that highlights that they are not complete. Every time a user engages with it, a map is completed. Google Maps is the quintessential Blind Map. It has a fundamentally emergent status: it looks empty and its colors are pale as if the map anticipates being filled with a highlighted location or route. Opening the app, the software does not show a map that is complete, but one that is the starting point of a process of searching, scrolling and zooming in and out. In essence, Google Maps is a processual map.

The map that Google's cartographic app shows on mobile devices is different for each user. Naturally, this has to do with the particular device that is used to access the software, the size and proportion of the display and the brightness settings of the screen. But more importantly, it is different because what is shown depends on the location where the software is used: the Google Maps app on mobile devices crops the map in such a way that the location of the user is placed in the center of the map. The user's position is shown as a small blue dot on the map. For me, this blue dot is a cartographic innovation of unprecedented magnitude, perhaps the biggest change in thinking about maps, mapping and mapmaking since The Blue Marble photo from 1972 made us look at Earth in a different way.[42]

Google has named the feature My Location but I will call it the Blue Dot, because it is much more than just a functionality that indicates one's position. I regard the Blue Dot as a symbol that highlights a shift in the practice and theory of mapmaking in which the role of the user has become more important. The user is not the recipient of a graphic product at the end of a process, but a co-creator. The Blue Dot literally puts the user on the map. It is both a visual sign indicating the presence of a user in a graphic product and an emblem marking a new phase in the thinking about production and use. The binary division between producer and user no longer applies here. Furthermore, what appeals to me in the name Blue Dot over My Location is that it is a concept that evokes a visual form. Hearing it is seeing it. Other concepts that have a similar quality are The Blue Banana[43] and The Duck.[44]

The Blue Dot is one of two items in Google's mobile cartographic app that remain the same when the map is zoomed in or out. All other elements change in size and color or appear or disappear when moving from close-up to long shot. The other item that is always there is the Google logo; the company name in green, red, yellow and blue letters has a fixed position in the left bottom corner. Google and the user both make the map. Both also use the map. The user to obtain information and Google to obtain data about what is acquired.

The Blue Dot is a cartographic innovation, especially from a post-representational point of view. The blinking circle underlines that Google Maps is processual and that the binary division between producer and user no longer applies. The user is as much the maker as the producer of the cartographic app as her location determines the crop of the map. Through scrolling and zooming she changes what the cartographic app shows, she can demand locations and routes to be displayed and she can add information to the map. The user not only controls it, she is also ever present in the map of Google. The map is always her map, her interpretation. The fact that she has a presence on the map confirms this.

At the same time that the functionality in Google Maps is empowering the user, she is also being used. Google records the user's location and movements. The data thus generated is the company's raw material that it processes and monetizes. The Blue Dot is animated, it expands and contracts in a regular rhythm, the effect of this movement is that it appears to be breathing. This breathing not only confirms the presence of the user, but also the presence of the company that is observing and recording. Blue Dot is watching you.

Another way that Google Maps harms the user is that it provides too little overview. The limited resolution and the small size of the screen of mobile devices does not give overview, which makes orientation difficult on a map scale other than showing the close proximity of the user's location. On larger scales, when more zoomed out, the map contains too little information, both in quantity—how much is shown, and in quality—how it is shown: the colors are too pale to easily distinguish the various cartographic parts.

There is a third way in which the Blue Dot user is a victim. Google Maps, in a way, is too complete, it offers a diagrammatic version of reality including the user herself. It does not leave room for additional, alternative versions of that reality. The process of orientation of a map user is comparing reality with an alternative version of that reality, simplified and diagrammatic. Google Maps, however, is not about comparing; the alternative version of reality it provides is complete in that it includes you, the user. Google Maps is the only version of reality. Occasionally you still see someone on the street holding a map. Alternately looking up and down, peering around in a dazed way and then scanning the map to get clarity. Orienting is comparing. More often you see someone in the street fixated

42 Taken on 7 December 1972 by the crew of the Apollo 17 spaceship on its way to the Moon, The Blue Marble image of Earth is one of the most reproduced images in history.

43 The Blue Banana is the name given to a curve shaped corridor of urbanization stretching from North Wales across Greater London to the Benelux and across the German Rhineland via Switzerland to Northern Italy. The color blue refers both to the color of the European flag as to the blue collars of the factory workers in the region. The Blue Banana spatial concept was coined in 1989 by RECLUS, a French group of geographers.

44 The Duck is a term to describe certain modern architecture that is expressive in form and volume as opposed to The Decorated Shed which characterizes a particular premodern architecture that relies on imagery and signs. These concepts were introduced in Venturi, Scott Brown and Izenour, *Learning from Las Vegas*.

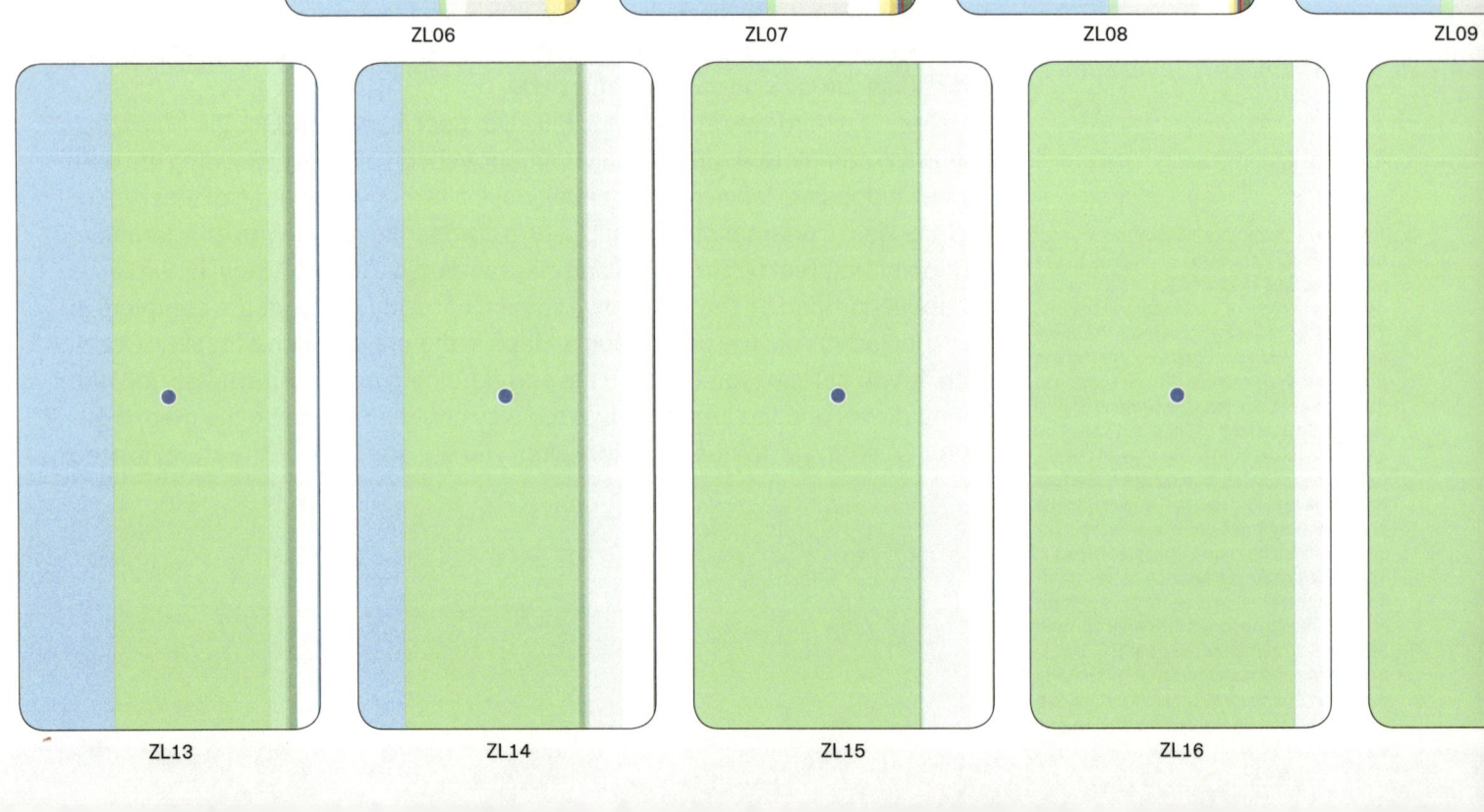

ZL00

ZL06

ZL07

ZL08

ZL09

ZL13

ZL14

ZL15

ZL16

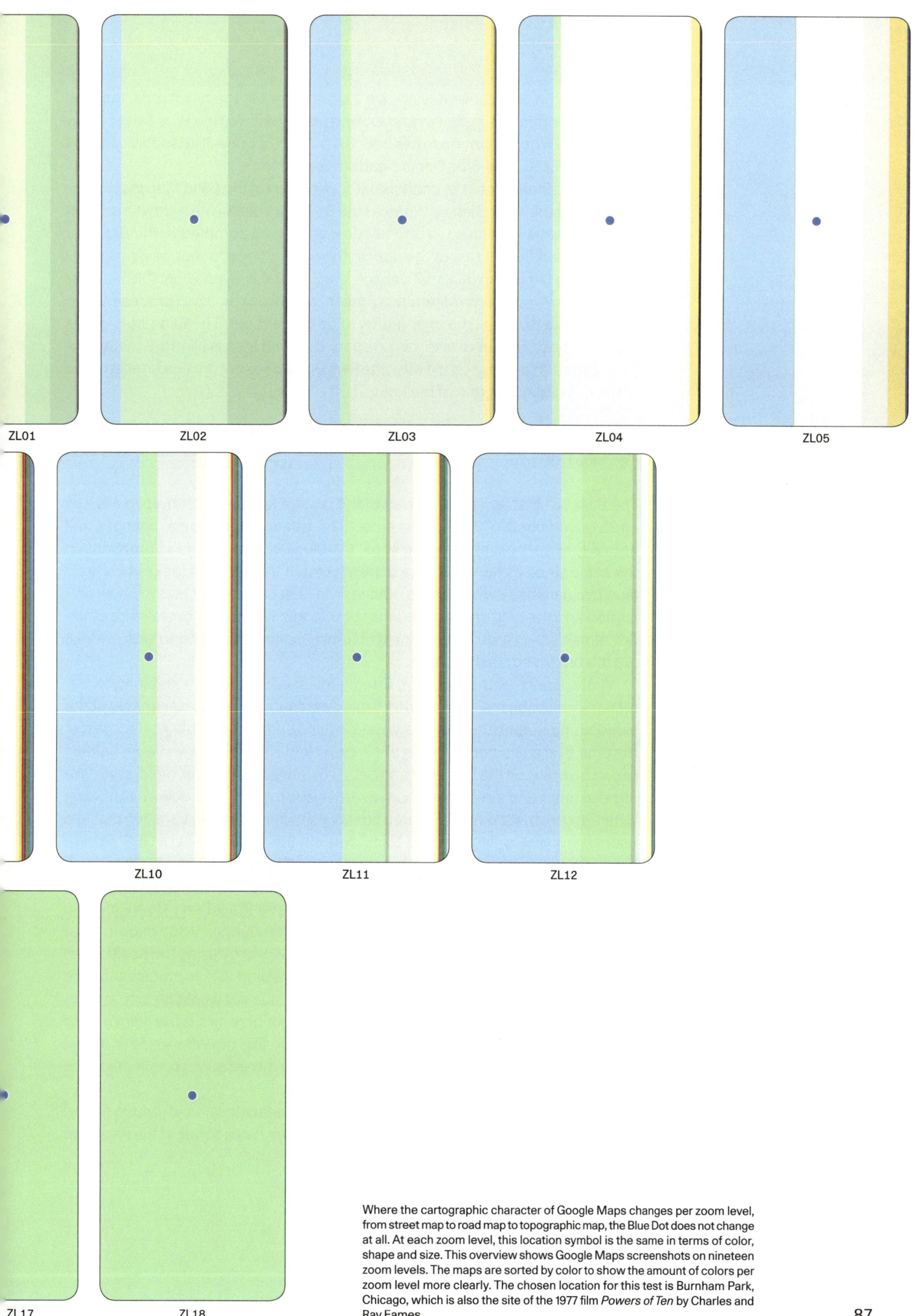

ZL01 ZL02 ZL03 ZL04 ZL05

ZL10 ZL11 ZL12

ZL17 ZL18

Where the cartographic character of Google Maps changes per zoom level, from street map to road map to topographic map, the Blue Dot does not change at all. At each zoom level, this location symbol is the same in terms of color, shape and size. This overview shows Google Maps screenshots on nineteen zoom levels. The maps are sorted by color to show the amount of colors per zoom level more clearly. The chosen location for this test is Burnham Park, Chicago, which is also the site of the 1977 film *Powers of Ten* by Charles and Ray Eames.

87

on the map on her mobile device, like a deer caught in the headlights, focused on an image that she is part of and looking at the beam of the Blue Dot to know what direction to go. Orienting has become looking at oneself in Google's filtered version of reality. The Blue Dot is a cartographic mirror.

I see the Blue Dot as the conceptual counterpart of the Blind Map that I introduced in the previous chapter. Whereas the Blind Map deals with a graphic product that is essentially unfinished and needs a user to be completed, the Blue Dot involves the different role of the user in the production process of visual information. The Blind Map deals with space, the Blue Dot with position. The Blue Dot gives the user, sometimes literally, a presence in the mapmaking process. But in order to take this role, the user needs to be given room. The Blind Map offers opportunity to the user to take this position. The Blind Map is blind in a metaphorical way as it is centered on the user, the Blue Dot; it does not provide an alternative version of reality, but that of the user.

The Blue Dot

The Blue Dot first appeared as part of the Google Maps app for mobile devices on 28 November 2007.[45] Given the amount of updates, restyling and changes that normally occur on websites and apps, the Blue Dot has remained pretty much the same since its launch. In the current version the Blue Dot looks like a solid blue circular shape with a white outline. The blue circle is not static but an animation and the only animated element in Google Maps. The overall size of the dot remains the same, but the inside blue shape grows and contracts. It looks like it is breathing, as if it is alive.

At the launch of the Blue Dot, Google released a two and a half minute animation explaining the new feature.[46] It is interesting to take a closer look at the video to understand the use Google had in mind and the technological context at the time. The animation starts by showing a clumsily hand-drawn figure holding a map, standing on the corner of a street. A voice-over (American, male) says: "We all need maps and directions when we are on the move. The problem is knowing where you are right now." The video continues to introduce My Location that lets users know where they are in "1-Click" without the requirement of GPS. Shown is a hand-drawn smartphone that resembles the BlackBerry Curve 8300 series.[47] The original 8300 model did not have a GPS antenna that enabled it to receive satellite signals to detect its location, although later BlackBerry Curve models would have GPS functionality. The My Location introduction video explains that Google Maps determines the location either via GPS or by using the positions of cell towers in the phone network. As the latter method is less precise than the GPS method, a larger light blue circle around the blue dot would be displayed to show the approximate location. The larger the light blue circle, the less precise the user's geographic position can be detected. To find out where one is, a user of Google Maps had to press the "0"-key of the keyboard and the animated blue circle would be visible on the map.

The video continues with an example. John (a faceless hand-drawn figure) has just arrived in London (a drawing of the Big Ben clock tower of the Palace of

Westminster is shown in the background) and is hungry. "John wants his first meal to be the London curry he heard so much about."[48] Rather than typing in the location on phone, "Tottenham Court Road," John writes "curry" and presses "0" to find his way to a local restaurant to eat chicken tikka masala. From the example in the animation it becomes clear that the goal of the My Location technology is to connect users and companies who provide goods or services. In that sense it fits in Google's economic model, the abovementioned surveillance capitalism.

After the explanation and example the video addresses the issue of privacy. "You might ask, does Google know where I am? The answer is No." The animation goes on to explain that Google uses the same information as telephone companies and that it only knows where a phone is, not who is using it, their phone number or any other information. "And if you want you can always disable the feature." Today we know, for example through the leak of Edward Snowden, that location information and Google Maps searches can be traced by others than the users.

Since its launch in 2007, the Blue Dot format did not fundamentally change. In 2016 a blue beam was added to the Blue Dot to indicate direction.[49] The main look and functionality of the Blue Dot remained the same as first presented in the introduction video. In fact, the Blue Dot has become the standard in several map apps. Apple's cartographic app Apple Maps as well as the Chinese language map app Baidu Maps both have a blue dot to indicate the user's location. And also the functionality—the central position on the map when opening the app, the larger-sized light blue circle to indicate a less precise location indication, the animated contraction and expansion of the circle, the unchanging size of the blue dot when zoomed out or in: all these aspects have been adopted by the cartographic apps of Google's competitors.

The fact that the Blue Dot has not fundamentally changed since its launch in 2007, that it is globally used by billions of users and that its look and functioning have been adopted by others, to me signifies that it has been well designed. I became curious to discover who designed the Blue Dot. I had the opportunity to find out when I met Jonathan Lee, a design manager at Google.[50] Both of us were speaking at AGI Open, a graphic design conference in Biel/Bienne, Switzerland, in 2015. Via e-mail, Lee put me in contact with Sanjay Mavinkurve, the designer of the Blue Dot. In an e-mail exchange I asked Mavinkurve about his motives behind the design. Why is the Blue Dot blue, and why does it remain the same size if the map changes scale? Mavinkurve is not trained as a designer but has a MSc in Computer Science from Harvard University. From 2003 to 2011 he worked for Google as User Experience Design Manager, currently he is Director of User Experience at Google Play.[51] The choice of the dot shape makes sense, according to Mavinkurve, as this is a common symbol on maps to indicate a location. In our exchange, Mavinkurve indicated that the limited capability of the smartphones at the time was the reason for the dot remaining the same size when the map changes in scale. He noted that other elements on the map, like stars and red location markers, also did not scale according to zoom level. "As for the color ... also no good reason except that, at the time, blue was the one color I would have most tied with Google."[52] Mavinkurve continues that blue was the dominant color in Google's user interfaces.

45 Chu, "New Magical Blue Circle on Your Map."
46 "Google Maps for Mobile with My Location (beta)."
47 BlackBerry Curve is a smartphone from the Canadian Research in Motion (RIM) Company first released on 10 May 2007. The BlackBerry Curve was aimed at professional users, had a small 320 × 240 pixel screen, a full Qwerty-keyboard and trackball.
48 Curry is a dish originating in southern India that uses a combination of spices. From the early nineteenth century curry restaurants were opened in London. The curry houses became increasingly popular in Britain due to the large number of British colonial servants and military personnel who returned from India. After World War II, the popularity increased even more due to the large number of immigrants from South Asia. And although the British Foreign Secretary Robin Cook in 2001 declared the chicken tikka masala to be a true British national dish that reflected the country's multicultural pluralism, I cannot see the "London curry" remark in the My Location video but in the light of the impact colonialism has on the world of technology today. To name two examples: the basis of the physical infrastructure of the Internet is formed by the submarine cable network for telegraph communication between the different parts of the British Empire. Besides, India plays an important role in the tech industry today. Many technology companies are partly based in India and several important employees of those companies are Indian-born, including Sundar Pichai, the current CEO of Google, and Sanjay Mavinkurve, the designer of the Blue Dot.
49 Buczkowski, "Google Maps Get Redesign of the Blue Dot Showing Your Position."
50 Jonathan Lee obtained his BFA in Communication Design/Graphic Design at Pratt Institute, New York. After working at various graphic design studios, including 2×4, the studio of Michael Rock who was mentioned in the previous chapter, he started working at Google in 2011. At Google he was responsible for, among other things, the creative direction of the Google rebrand in 2015. Since 2017 Lee has been a critic at the graphic design department of the Yale School of Art. "LinkedIn profile Jonathan Lee."
51 "LinkedIn profile Sanjay Mavinkurve."
52 Sanjay Mavinkurve, email message to author, 18 November 2015.

1   09.2006

2   07.2008

3   09.2009

4   06.2010

8   09.2014

9   04.2015

10   09.2016

This chronological overview shows smartphones from 2007, the year the Blue Dot was introduced, to the present day. From every year one of the most sold phones has been chosen. The phones' screens are colored to indicate the resolution of the screen, the darker the color, the higher the degree of detail visible on the screen of the phone. In the course of time the screens of smartphones could not only display more information because they became larger, but also because the screen resolution became finer. The dot on each phone indicates the size of the Blue Dot in the Google Maps application.

Resolution (ppi)

150   200   250   300   350   400   450   500   550   1 cm

5 05.2011

6 09.2012

7 04.2013

11 04.2017

12 09.2018

13 03.2019

1	BlackBerry Pearl 8100	157 ppi	8	iPhone 6 Plus	401 ppi
2	iPhone 3G	163 ppi	9	Samsung Galaxy S6	577 ppi
3	Nokia 5230	229 ppi	10	iPhone 7 Plus	401 ppi
4	iPhone 4	326 ppi	11	Samsung Galaxy S8+	529 ppi
5	Samsung Galaxy S II	217 ppi	12	iPhone Xs Max	458 ppi
6	iPhone 5	326 ppi	13	Huawei P30 Pro	398 ppi
7	Samsung Galaxy S4	441 ppi			

I understand Mavinkurve's color choice. Blue is the predominant color on social media. Nearly all logos of social media companies use the shade.[53] But given the shape I thought that the Blue Dot was referencing "The Blue Marble" or "The Pale Blue Dot," respectively: the ubiquitous photograph of Earth taken on 7 December 1972 by the crew of the Apollo 17 spacecraft at a distance of about 45,000 kilometers and a photograph of planet Earth taken on 14 February 1990 by the Voyager 1 space probe from a distance of about 6 billion kilometers. In the latter photograph, Earth's apparent size is less than a pixel; the planet appears as a tiny dot against the vastness of space, among bands of sunlight scattered by the camera's optics. But while my thoughts wandered off to remote distances and far away perspectives, the reality of the Blue Dot is that it is about this place at this moment in time. The Blue Dot is a cartography of the here and now. The blue symbol suits a world where through information technology we have more connections with, and knowledge of, the rest of the world. The animated dot that appears to be breathing is a fitting symbol that seems to say: I am the Blue Dot, I am right here right now, I am.

I admire the design of the Blue Dot. But, as with other designs that I respect, I sometimes wonder how I would have designed it myself. In the case of the Blue Dot there are two nagging aspects that I would like to revisit. First it is the shape of the dot. As it is a solid form it covers the area where the user is located. When completely zoomed in this is not a problem. But when zoomed out a bit, the dot is relatively big and can cover complete buildings or side streets. An open or outline shape could solve this. But for this open form to be visible enough it would need to be so big, or the outline so thick, that it would have too big a presence on the map. The other aspect I would like to reconsider is the immutability of the Blue Dot throughout the various scales. With each zoom in or out, the map of Google changes character, from a street map at closer zooms to a road map to a topographic map in views from further distances. In the different scales the relation between the user and the map changes, one would expect the symbol of the user's location to reflect that. In a close view one would almost expect to see oneself, or at least the Blue Dot to have more detail. In views from far away the dot almost becomes ridiculously big and important when it is covering cities or even countries. Why could the Blue Dot at this global scale not be smaller, perhaps reduced to a mere pixel?

Conclusion

In this chapter I looked at different modes of thinking about maps. A cognitive, functional approach built on the premise that the world can be known and truthfully mapped; a critical approach that sees maps as complex and contested, as products and producers of power and a processual approach that questions the ontological foundation of the map and regards the map as mapping, as a process that is never complete. Each of these modes of thinking generates considerations that become criteria to evaluate the look and functioning of a map. Conversely, how the design of a map is perceived is informed by the viewer's conceptual understanding of the cartographic product. From this it follows that a purely functional reading of the map without it being questioned conceptually does not suffice. Nor does a critical reading of the map solely focused on understanding the hidden structures suffice, as it does not provide criteria for how a map should function. Building on this idea, it can be said that design invites theory. And theoretical understanding of a product is necessary to understand its design. In mapmaking, theory and design are intertwined.

This chapter looked into the question of whether it is possible to produce fundamentally new maps, and if so how. Following the above it can be said that a map that looks and functions differently will only be perceived as innovative if the new design is linked to a different conceptual understanding. A full understanding of the map would need to scrutinize aspects like how the map is produced, by whom, for what reasons, employing what tools and how is it used.

Post-representational cartography perceives the map as a process. Google Maps is the quintessential example of a processual map, because it is in a constant state of becoming. Users search, scroll and zoom to make the map complete. A processual reading of maps questions the producer–user divide and in Google Maps it certainly is the case that the user is as much a producer of the maps the app displays as Google is itself. At the same time, the Google Maps user is a victim as her data is used as raw material exploited by Google. On another level she is prey because orientation using Google Maps is not a process of comparing, but about seeing a single version of that reality that is blind for alternative versions other than that of Google.

I introduced two concepts for this processual approach to maps and mapmaking. The Blind Map describes the emergent status of maps and other visualizations. Maps are never fully formed but are completed every time a user engages with it. The Blue Dot is both a visual sign indicating the presence of the user on a map and it is an emblem marking a different phase in the thinking about production and use in which the binary division between producer and user no longer applies. The two concepts are interrelated. The Blue Dot gives the user a presence in the mapmaking process. But in order to take this role, she needs to be given the opportunity to do so. The Blind Map offers space for the user to take this position.

53 On 30 April 2019 Facebook CEO Marc Zuckerberg unveiled a redesign of the Facebook mobile app and website discarding the color blue. The new design marked a shift of direction of the company to focus more on private messaging and less on public communication. The reorientation of Facebook came after accusations that the social network was used as a tool for election interference, that it spread false news, and that it did not properly protect the data of its users. In the "Fade to White" episode of *The Observatory* podcast, American graphic designer, educator and author Michael Bierut named the elimination of the color blue a de-branding strategy.

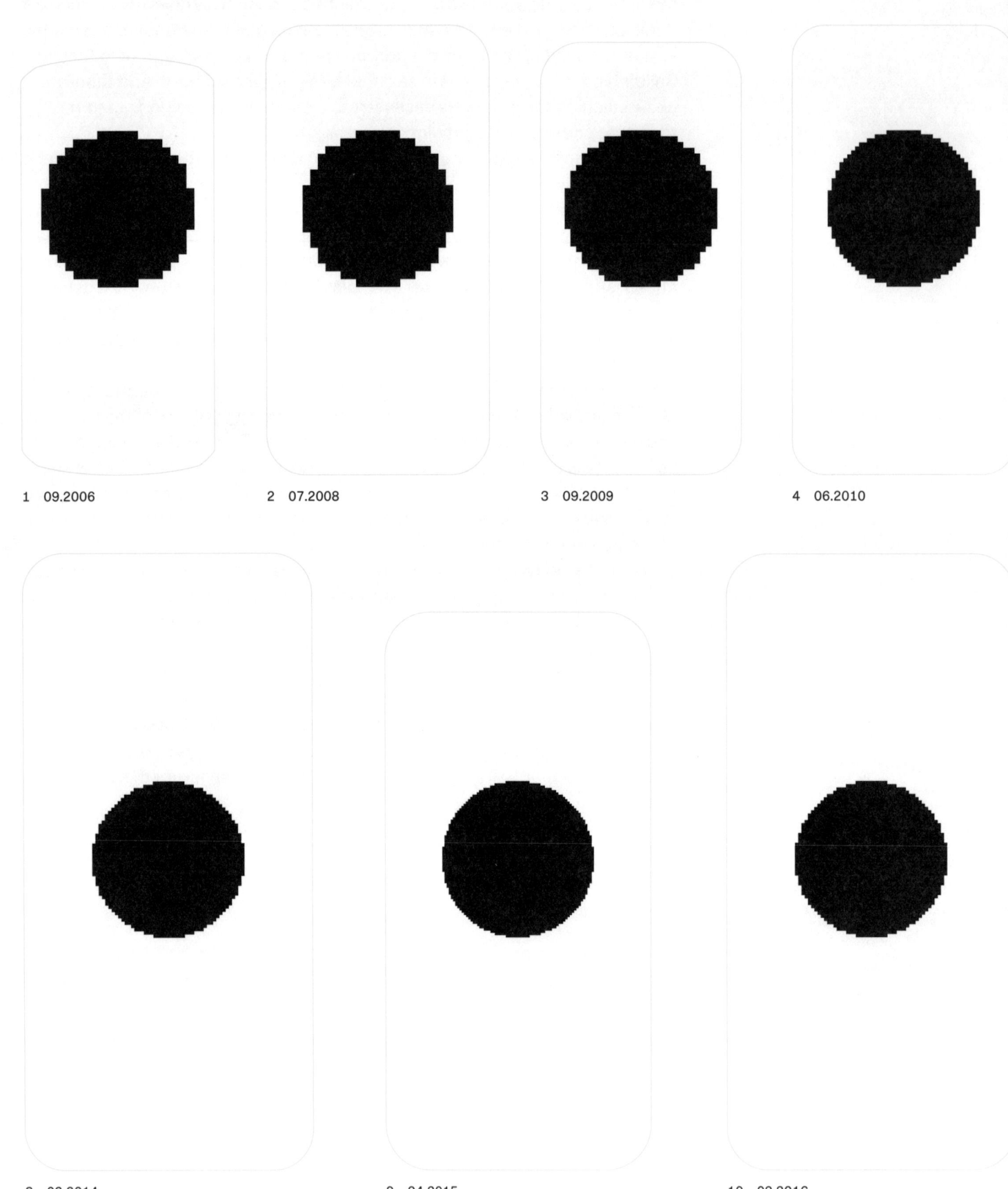

1   09.2006   2   07.2008   3   09.2009   4   06.2010

8   09.2014   9   04.2015   10   09.2016

The overview shows how the shape of the Blue Dot changed over time and became rounder due to increasing screen resolutions of smartphones. The size of the Blue Dot has been magnified ten times so that the changes are more clearly visible.

5 05.2011   6 09.2012   7 04.2013

11 04.2017   12 09.2018   13 03.2019

1	BlackBerry Pearl 8100	157 ppi	8	iPhone 6 Plus	401 ppi
2	iPhone 3G	163 ppi	9	Samsung Galaxy S6	577 ppi
3	Nokia 5230	229 ppi	10	iPhone 7 Plus	401 ppi
4	iPhone 4	326 ppi	11	Samsung Galaxy S8+	529 ppi
5	Samsung Galaxy S II	217 ppi	12	iPhone Xs Max	458 ppi
6	iPhone 5	326 ppi	13	Huawei P30 Pro	398 ppi
7	Samsung Galaxy S4	441 ppi			

4   The Strava Global Heatmap

The Strava Global Heatmap

In this chapter I will investigate the Strava Global Heatmap, a map produced by a social network for athletes called Strava, which is based on data generated by the users of its fitness tracking app.[1] Although it looks like a regular map with lines in different colors on a base map indicating water and relief, I will argue that the Strava Global Heatmap is not as ordinary as it may seem at first glance. There is a disconnection between what you see and what is on the map. It raises questions about what it shows, how it is made and whether it can count as a map at all. By identifying, untangling and scrutinizing technological, economic, social and cultural aspects of the map, and using knowledge from different fields, I will try to gain insight into the complexities of the Strava Global Heatmap.

The Global Positioning System (GPS) is a technology that is essential for all three mapmaking practices investigated in this book: Google Maps' Blue Dot, the Strava Global Heatmap and Thomas van Linge's map of Syria. The satellite-based navigation system, initially developed for military strategy purposes and later opened up for civilian use, is the key technology used to determine a user's position and plot it as a blue dot on the map app on her mobile device (Chapter 3). And it is GPS that is used to define the location information that is inserted as so-called geotags with digital photographs that form the raw material for amateur conflict mapmakers to produce their maps (Chapter 5). Unlike the chapters on the practices of Google Maps and Thomas van Linge, this chapter will also look into the infrastructure behind GPS and the hardware and service industry it prompted.

The Map

Published in November 2017, the Strava Global Heatmap shows in detail the aggregated, public activities of users of the Strava fitness tracking app of the last two years.[2] The Strava app enables users to record their physical exercise, but it is also a social media platform for sharing these activities. A heat map is a visualization of data where values are represented by a range of colors: the higher the value, the "hotter" (brighter and/or warmer) the colors. A heat map is not necessarily a map, a geographic visualization, it can also be a diagram depicting non-geographic data. The "heat" in the Strava Global Heatmap is the rendering of the activities of Strava users as lines on a world map. The more a certain route is run, cycled or swum, the brighter it will be displayed on the map. The Strava Global Heatmap became controversial when in January 2018 a military analyst discovered that the map shows the location of secret military bases.[3] Ironically, GPS—the technology that lies at the basis of fitness apps like Strava—was originally a closed-off military technology that was later made available for civilian use, partly to support the developing US commercial GPS equipment and service industry. After the discovery of the sensitive information about the locations of secret military bases, the discussions in the news media focused on the privacy of users of social media and the power of technology companies like Strava, Facebook and Google to track our behavior and use it to control our lives.

Privacy will not be the sole focus here. The main question is not so much what is revealed on the map, but what do I see when I look at the Strava Global Heatmap? What does the Global Heatmap reveal about Strava? The visual strategies of the

1   The fitness app and the heat map are two separate entities that share certain aspects: the fitness app contains maps and the heatmap is a piece of software that the user can interact with. In this chapter I will use the term *app* for the Strava fitness app and *map* for the Strava Global Heatmap.
2   "Strava Global Heatmap."
3   Hern, "Fitness Tracking App Gives away Location of Secret US Army Bases."

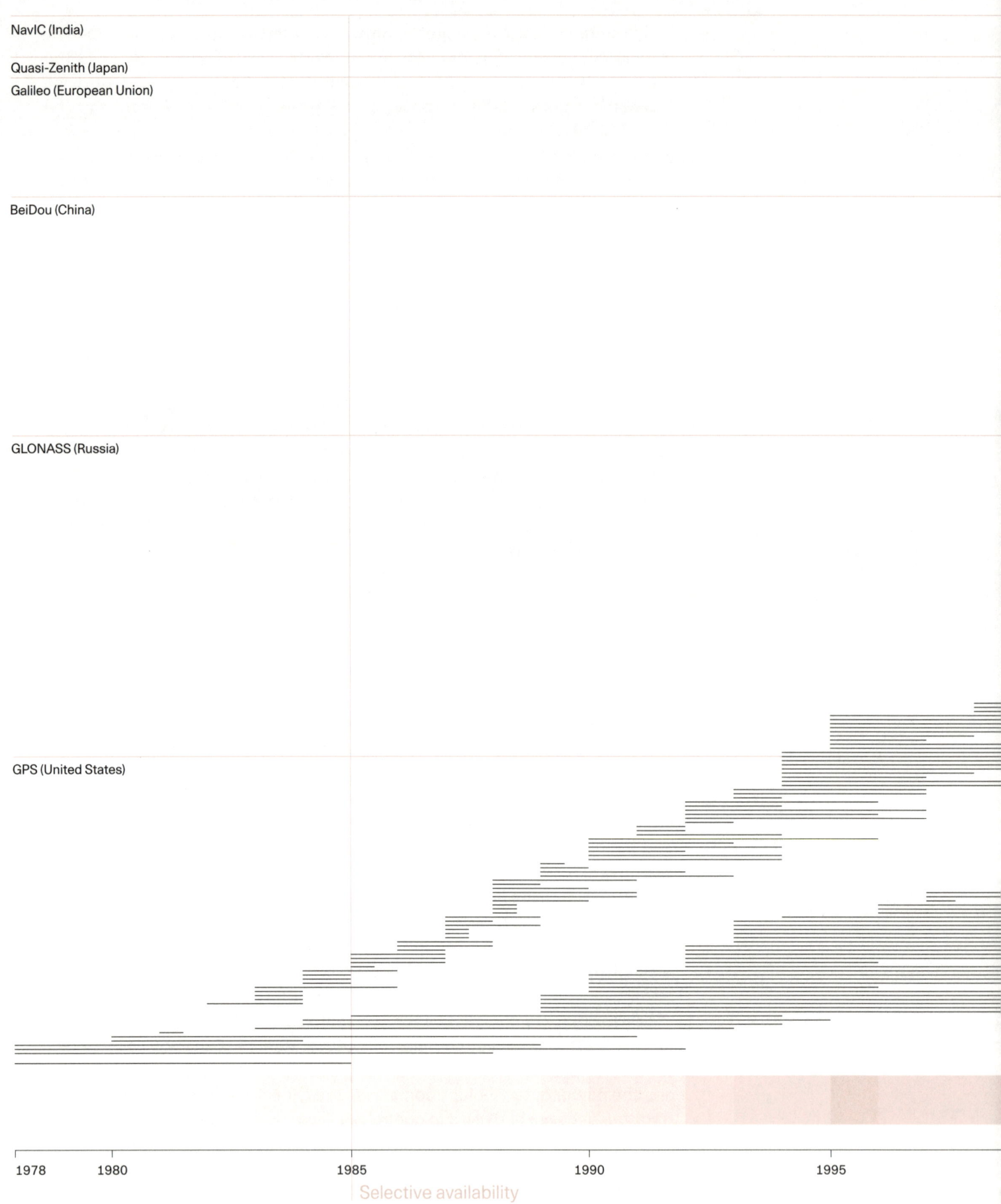

NavIC (India)

Quasi-Zenith (Japan)

Galileo (European Union)

BeiDou (China)

GLONASS (Russia)

GPS (United States)

1978   1980   1985   1990   1995

Selective availability

This overview shows the number of active navigation satellites, both from the Global Positioning System (GPS) and from other systems. Each horizontal line represents the time a satellite is active. Satellite navigation systems consist of multiple satellites: the GLONASS network, for instance, consists of 24 satellites. Most smartphones support multiple navigation systems. The more satellites a smartphone can receive, the more accurately it can determine its position.

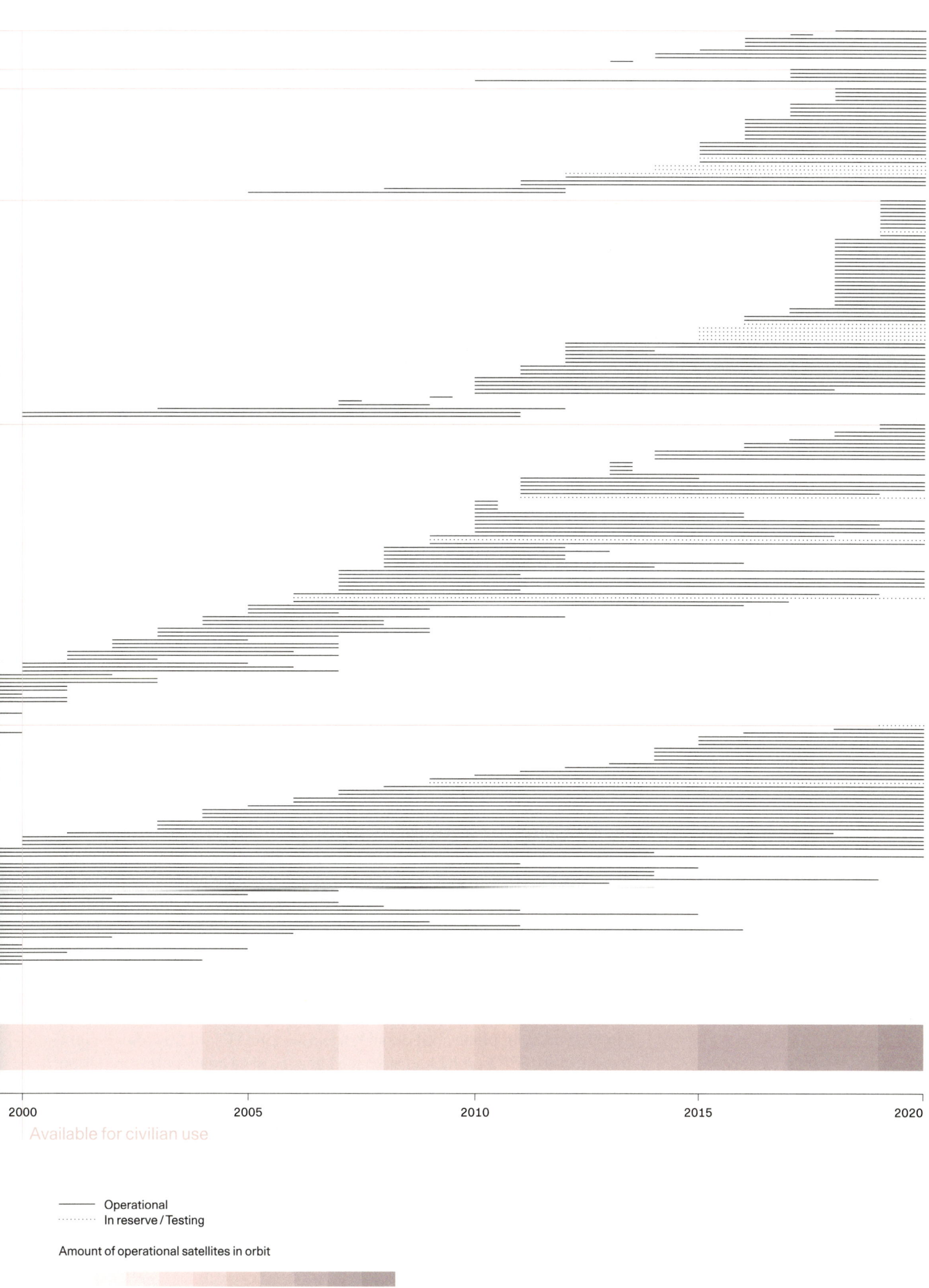

Available for civilian use

——— Operational
·············· In reserve / Testing

Amount of operational satellites in orbit

10   20   30   40   50   60   70   80   90   100

map with its glowing lines are seductive, but I suspect there is more beneath the surface. I will look critically at a diverse range of issues, like the technological, sociological, economic, cultural and surveillance aspects behind the Strava map. Others have linked these fields to analyze current digital phenomena. My approach is to take a concrete example, the Heatmap, and link considerations from a variety of fields to its design—how it communicates and how it is produced—and the complex relationship between its producer and user.

A second topic of this chapter will be the impact of GPS on the field of cartography and how this can be explicitly observed in the Strava Global Heatmap. This second focus links this case study to the overall topic of this book: the transformation of the field of graphic design as a result of technological developments of tools to create, record, edit, produce and distribute visual information, and their influence on the positions of producers and users.

The third focus is methodological, referring to the field of artistic research. Often projects in this particular field result in two parallel but separate kinds of output: academic and artistic. It is my aim to merge these two flows in formats or approaches that belong to both of these fields. For this chapter, the timeline serves as such a medium. The reason is that the Strava Global Heatmap can be much better understood by looking at temporal aspects like developments of technologies of ideas, or the linking of events, than by looking at a static object. This third topic has an ambition beyond the scope of the investigation: how to reshape my artistic practice into a research practice that uses graphic design as a means to conduct research and to disseminate the findings.

Satellites

GPS, originally Navigation System Using Timing and Ranging (Navstar), is a network of satellites and ground stations that provides users carrying a portable receiver with precise information on their location, anywhere, anytime and under any weather circumstances.[4] GPS satellites carry atomic clocks that are synchronized with one another and with ground stations. Thirty-two satellites orbit in six circular paths around the Earth at an altitude of 20,200 kilometers, transmitting data about their current time and position. A GPS receiver monitors multiple satellites and computes its precise position by measuring the time it takes to receive the different signals. At a minimum, four satellites must be in "view" of the receiver to calculate its three-dimensional location (latitude, longitude and altitude).

The satellites, owned by the US government and launched and operated by the US Air Force, emit their signals to anyone with a GPS receiver, including those embedded in cameras, smartphones and smart wearables like fitness trackers. GPS was initially developed as a military technology: the first of its satellites was launched in 1978. For reasons of safety and economy the system, over the course of time, was opened up to civilian use.

In 1983 a Soviet plane shot down Korean Air Lines Flight 007 after it went astray into the USSR's prohibited airspace because of navigational errors, killing all 269 passengers. Following this incident, to promote increased safety for civil aviation, the Reagan administration announced that GPS would be made available as a "dual-use technology" once it was completed.[5] This meant that GPS would be available to everyone, but that the signal for civilian use was intentionally downgraded. Whereas the military signal would provide an accuracy of up to five meters, the civilian signal would give a position with an accuracy of up to 100 meters. It was, however, possible to improve the accuracy of this downgraded signal using a technique called "differential correction," which involved gathering additional readings from base stations at known locations within roughly 500 kilometers (the area covered by four satellites). This enabled the reading of a position with an accuracy between two and five meters. Differential correction made the civilian signal close to or as good as the military signal and prompted the development of navigational hardware for hand-held and automotive use.

The decision to end the "selective availability" of the signal was then made in 1996 by the Clinton administration, following a recommendation by the Secretary of Defense. From 1 May 2000 the same GPS signal became available to all. The reasons for this decision were twofold. The US military had developed technologies to deny GPS to potential adversaries in areas of operations while preserving the peaceful use of GPS services outside those areas. Also, the US government acknowledged that GPS had become an integral part of the global information infrastructure that had generated a US commercial GPS equipment and service industry that was a leading world player.[6]

GPS was first used on a large scale in the Gulf War (1990–1991)—also known as Operation Desert Shield for the build-up of troops and defense of Saudi Arabia, and Operation Desert Storm for the combat phase—in which a US-led coalition expelled Iraq's forces from Kuwait. With only sixteen of the planned twenty-four satellites in orbit, a number that would not be reached until April 1995, the system provided a coverage that lasted approximately nineteen hours per day.[7] GPS provided troops with accurate information about their location in the desert at day and during the night, which is seen to have contributed considerably to the swift 100-hour ground campaign. In contrast to the ubiquity of GPS technology today, the number of GPS receivers the US Army had at its disposal at the outset of Operation Desert Shield was limited. Of the 40,000 vehicles used in the battle, only 3,000 were outfitted with a GPS receiver. Most of these units were commercially produced models. Seven different types of GPS devices were used during the Gulf War. The two most frequently used models were the AN/PSN-10 Small Lightweight GPS Receiver (although at nearly four kilograms the term *lightweight* is relative), produced by Trimble Navigation, intended for use in vehicles and aircraft; and the smaller NAV 1000s, produced by Magellan from 1989, the first handheld consumer GPS device. The NAV 1000s sold at the time for $3,000 and measured 22.2 × 8.9 × 6.3 cm, weighed nearly 700 grams and were powered by six AA batteries, which allowed them to run for a few hours.

Following GPS, other global navigation satellite systems were built. The first satellite of Russia's Global Navigation Satellite System (GLONASS) was launched in 1982. GLONASS consists of 24 satellites and is operational since 1995. Named after Italian astronomer Galileo Galilei, Galileo is the satellite network of the European Union that will consist of 30 satellites once it is fully operational in the late 2020s. The first Galileo satellite was launched in 2011. The BeiDou Navigation Satellite System is the Chinese satellite network, which achieved full global coverage in 2020. The first BeiDou satellite was launched in 2000. Besides these global networks, some local networks are active, like NavIC (India) and QZSS (Japan),

4   Kurgan, *Close Up at a Distance: Mapping, Technology, and Politics*, 39.
5   United States Department of Commerce, National Oceanic and Atmospheric Administration, "GPS & Selective Availability Q&A."
6   United States Office of Science and Technology Policy, National Security Council, "Press Release US Global Positioning System Policy."
7   Dissinger, "GPS Goes to War: The Global Positioning System in Operation Desert Storm."

Politics

Soviet Union begins flight tests of the GLONASS satellite navigation system

US launches first GPS satellite. Restricted to military use only　　Fall of the Berlin Wall　　　　Fully operational const

Selective availability. Next to a high-quality signal for military use, a degraded GPS s

Conflict

Korean Air Lines Flight 007 shot down by Soviet military after navigation error by pilots

The Gulf War is the first military conflict to widely use

Yugoslav Wars

US employs Preda

Technology

Magellen NAV 1000, first commercial handheld GPS receiv

General Atomics RQ-1 Preda

World Wide Web developed at CERN, Geneva (Switz

Economy

GPS-based technology company Navman founded (New Zealand)

GPS-based technology company Garmin founded (US)

GPS-based technology company TomTom fou

Google fou

| 1978 | 1980 | 1985 | 1990 | 1995 |

Selective availability

This timeline compares the development of the number of navigation satellites (4.1) with developments in technological, political and economic power.

Number of monthly confirmed US drone attacks
in Afghanisatan, Pakistan, Somalia and Yemen
(Source: The Bureau of Investigative Journalism)
1,119

China launches first satellite of the BeiDou Navigation Satellite System

India launches first satellite of the NavIC Navig

:ion of 24 GPS and 24 GLONASS satellites

Japan launches first satellite of the Quasi-Zenith Satellite System

l becomes available for civilian use

European Union launches first satellite of the Galileo Glob

Selective availability discontinued. The same GPS signal for military and civilian use

partially denies GPS access to Indian military during the Kargil War

North Korea jams GPS US signals in South Korea

Russia jams GPS US drones in Ukraine

September 11 attacks

Malaysia Airlines Flight 17 shot down o

PS   War on Terror

First kill by a US Predator drone in Kandahar (Afghanistan)

Keyhole EarthViewer used by CNN during the invasion of Iraq sparks interest in Keyhole

ones for reconnaissance missions over Former Yugoslavia

S)

Google Street View

Google stock price
$ 1,503.21

notely piloted reconnaissance aircraft (US)   Google Maps launches "My Location" feature, the Blue Dot

MQ-1 Predator, armed version of RQ-1 drone (US)   Google Earth reaches 1 billion downloads

d)

Google releases EarthViewer as Google Earth

Google releases Google Maps, the cartographic overlay of Google Earth

Geospatial data visualization firm Keyhole founded (US)   Google Maps most popular app for smartphon

Central Intelligence Agency invests in Keyhole   Google has over 7,100 employees working in mapping

(Netherlands)   Google acquires Keyhole   Over 3,000 navigation applications available in Apple's App Store

by Larry Page and Sergey Brin (US)   Google Maps appears on a smartphone for the first time

Garmin stock price
$ 102.41

Dot-com bubble caused by excessive speculation in internet-related companies

Tomtom stock price
$ 10.49

2000   2005   2010   2015   2020

Available for civilian use

103

that do not provide global coverage. All of these networks send out signals that can be received by newer devices. Sometimes up to twenty signals can be received, dramatically optimizing the accuracy of the determination of a user's location.[8]

Over time, GPS receivers became smaller, lighter, cheaper and more widely available. GPS tracking also became a feature on other devices such as digital cameras, adding location coordinates to digital photographs, to mobile phones, to smart phones, where GPS tracking is a standard feature, and to wearable technology like fitness trackers and smartwatches. The devices in the latter category are in close contact with the user and are equipped with additional technologies to collect data about the number of steps walked, calories burned, blood pressure and heart rate. In the fitness trackers the signals from remote satellites are combined with the user's internal signal of the heartbeat. GPS in these devices is not only used as a technology to know one's location, but also to track one's time.

GPS, together with the Internet, mobile mapping applications, geotagging and "open source" collaborative tools, is one of the technologies that "undisciplined" cartography.[9] The navigation system enabled nonspecialists to collect spatial data and map them out. And while the technology offered individuals possibilities to compile their "personal plots," as the GPS tracks are called by Lisa Parks, Professor of Comparative Media Studies and Science, Technology, & Society at the Massachusetts Institute of Technology, the Global Positioning System also provided opportunities for the state and commercial enterprises in their quest for "total knowledge."[10] The simultaneous difference between the "intimate particularity"[11] of the individual GPS track and the "total vision"[12] of an enterprise is one of the many dichotomies I encountered in the Strava case study.

Search Engines

The beginning of the twenty-first century saw an important step in the evolution of the World Wide Web, with users moving away from being only the consumers of content. The Web 2.0 emphasized user-generated content, ease of use, and the connection between the Internet and other products, systems and devices. Social media platforms like Facebook, YouTube, WhatsApp, Instagram and Twitter and web search engines like Google, Bing, Baidu and Yahoo! made it possible for users to express themselves, learn, connect and search at will.

Observing the implications of this next step in information technology, American social scientist and professor emirata at Harvard Business school Shoshana Zuboff sees a new economic and social logic that she calls "surveillance capitalism," in which the user's behavioral data are the source of new economic value.[13] Surveillance capitalism is the feedback loop between surveillance, the accumulation of information, and capitalism, the accumulation of wealth. It is a business model for companies to track, store and analyze everything of their users and monetize this by selling it to their customers.

Surveillance capitalism was invented by Google. After the dot-com collapse in 2001, Google was under pressure from its investors to make a profit. It tried to increase the revenue of advertisements by using the behavioral data that were stored but so far never used. Up until then the search terms that trigger the display of advertisements were selected by advertisers themselves. Using the substantial analytical capabilities Google had developed for its search engines on the vast amounts of behavioral data it had collected turned out to be a historic turning point for the company. Google had a substantial success in pairing advertisements to pages and making profit from the zero-cost asset behavioral data. In this new model, users were no longer an end-in-themselves, they became a means to make financial gains.

This business model made Google one of the wealthiest and most powerful tech companies in the world. The wealth it obtained allowed it to buy and develop further products, services and tools, like YouTube and Android, currently the operating systems used on 82 percent of all smartphones. Social media company Facebook, with over 2.4 billion monthly active users worldwide,[14] also adopted surveillance capitalism and with the profits it made acquired other companies like Instagram and WhatsApp, increasing its power.

In the 1980s the aforementioned Zuboff formulated three laws of surveillance.[15] First, everything that can be automated will be automated. Second, everything that can be informed will be informed.[16] And third, every digital application that can be used for surveillance and control will be used for surveillance and control, irrespective of its original intention. It is especially this third law that comes into play in surveillance capitalism.

Turkish/French developer and cyborg rights activist Aral Balkan has called surveillance capitalism "people farming," a business model that is transforming the public sphere: "Google and Facebook want us to believe they are parks but they are shopping malls."[17] Social media and the Internet are private space masquerading as public space. According to Balkan, the lack of public space in the digital realm is detrimental to democracy and lies at the heart of symptoms like fake news and hate and bullying on social platforms. Balkan calls the exponential power distance between users and the technology companies "Slavery 2.0" and claims that the technology companies care more about the profile—the digital you, than the user—the you.

David Lyon, social scientist and director of the Surveillance Studies Centre of Queen's University Kingston, Canada, sees in today's world of software and networks a reduction of the body to data and the creation of data-doubles on which life-chances[18] and choices hang more significantly than on real lives and their stories.[19] Lyon refers to Zygmunt Bauman's work on the "liquidity of modernity," introducing the term *liquid surveillance* to describe the ubiquitous technologies and new surveillance practices that constantly check, monitor, test, assess, value and judge us. The concept I take from Lyon's theories is the dissection of data and body in the surveillance practices of social media companies: disembodied data. Whereas in the case of Strava it is the body that generates the data: no body, no data.

Self-Tracking

Self-tracking is the recording and monitoring of specific features of one's own life. It is also called life-logging, personal analytics, personal informatics and the quantified self.[20] These practices go back to the keeping of a diary, but have changed with the dispersion of mobile digital devices. Activity tracking or fitness

8   "What is GPS."
9   Crampton and Krygier, "An Introduction to Critical Cartography," 12.
10 Parks, "Plotting the Personal: Global Positioning Satellites and Interactive Media."
11 Ibid., 218.
12 Ibid., 220.
13 Zuboff, "Google as a Fortune Teller: The Secrets of Surveillance Capitalism."
14 Clement, "Number of Monthly Active Facebook Users Worldwide as of 2nd Quarter 2019."
15 Zuboff, "The Surveillance Paradigm: Be the Friction—Our Response to the New Lords of the Ring."
16 Informating is the process of translating descriptions and measurements of activities, events and objects into information. It is a term coined in Zuboff, *In the Age of the Smart Machine: The Future of Work and Power.*
17 "Digital Dystopia: Tech Slavery and the Death of Privacy."
18 Life-chances, from the German *Lebenschancen*, is a concept of German sociologist Max Weber on the opportunities individuals have to improve their lives.
19 Lyon, "Liquid Surveillance: The Contribution of Zygmunt Bauman to Surveillance Studies," 325.
20 Lupton, "Self-Tracking Modes: Reflexive Self-Monitoring and Data Practices."

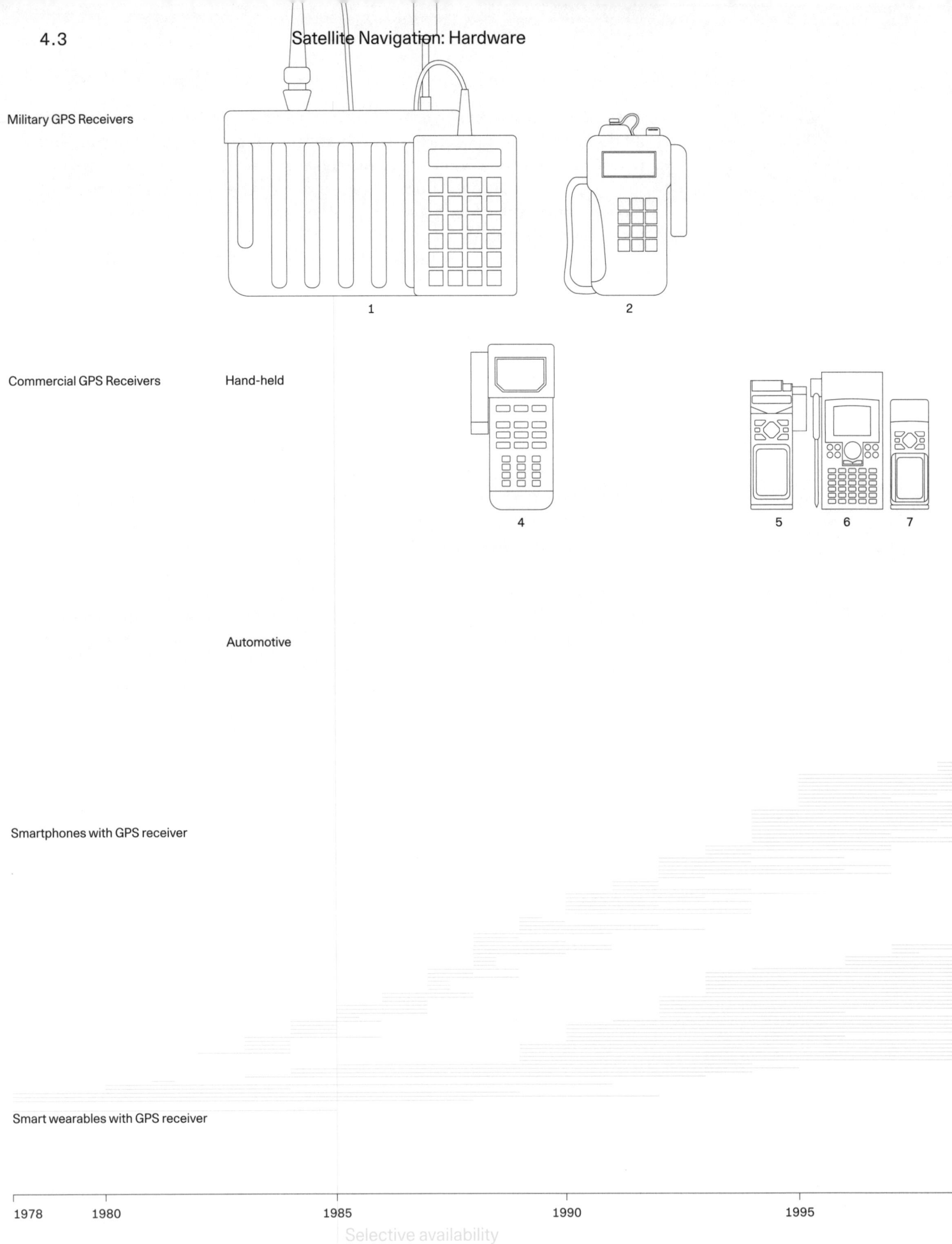

Military GPS Receivers

Commercial GPS Receivers   Hand-held

Automotive

Smartphones with GPS receiver

Smart wearables with GPS receiver

1978   1980   1985   1990   1995

Selective availability

This timeline compares devices that can receive satellite navigation signals,
from satellite navigation devices (also called GPS receivers), to car navigation
systems, to smartphones, to smart watches.

1	AN/PSN-8 Manpack GPS receiver	4	Magellan NAV 1000	17	Garmin StreetPilot	23	Benefon Esc!	36	Nike+ Sport GPS
2	AN/PSN-11 Precision Lightweight GPS Receiver (PLGR)	5	Garmin GPS 45XL	18	Garmin StreetPilot 2660	24	Blackberry 7750	37	FitBit Flex
		6	Magellan GSC 100	19	Garmin c340	25	iPhone 3G	38	SONY Smartwatch 3
		7	Garmin 12XL	20	TomTom IQ routes	26	HTC G1	39	Fitbit Surge
3	DAGR—Defense Advanced GPS Receiver	8	Magellan GPS 315	21	Garmin nüvi 3597 LMTHD	27	ZTE 945	40	SamsungGear S3
		9	Magellan SporTrak Map GPS	22	Magellan TRX7	28	Samsung Galaxy S	41	HuaweiBand 2 pro
		10	Magellan Meridian Gold			29	iPhone 4s	42	Fitbit Ionic
		11	Magellan exPlorist 100			30	iPhone 5s	43	Apple Watch Series 4
		12	Garmin eTrex			31	iPhone 6s	44	Garmin 945
		13	Magellan Triton 2000			32	SONY Xperia Z5		
		14	Magellan exPlorist 710 hiking			33	iPhone x		
		15	Garmin GPSmap 66s			34	iPhone xs max		
		16	Garmin edge explore			35	Samsung Galaxy S10		

3

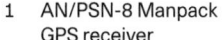

8   9   10   11   12   13   14   15   16

17   18   19   20   21   22

24   25   26   27   28   29   30   31   32   33   34   35

36   37   38   39   40   41   42   43   44

2000   2005   2010   2015   2020

Available for civilian use

107

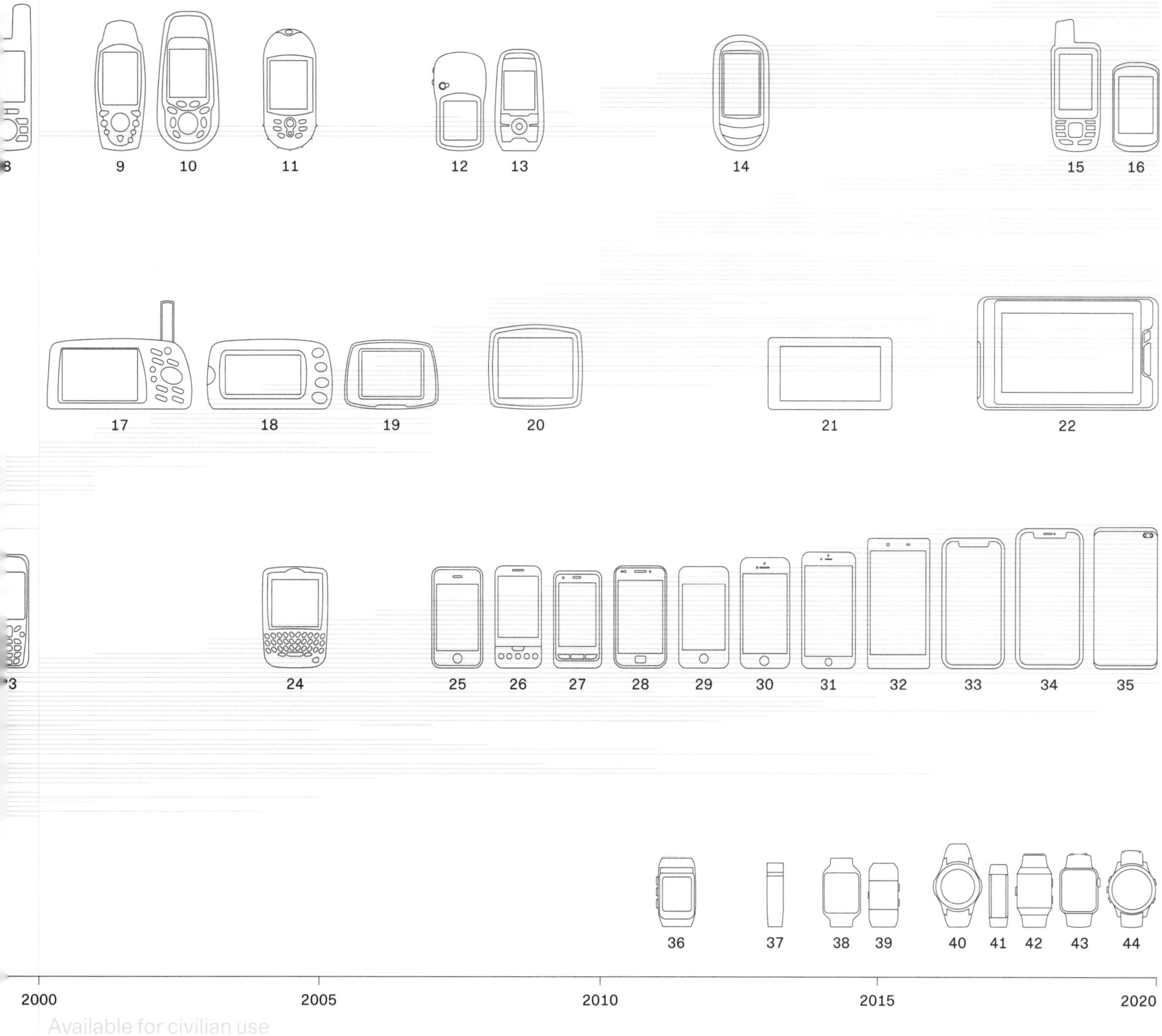

tracking is a specific kind of self-tracking. Using the GPS and other functionalities of mobile and smart devices, the position and time of sport activities is recorded as well as the condition of the body in terms of heart rate and blood pressure. Apps keep track of the recorded data and combine these to give users insight and comparison.

Strava is a social network for athletes, more specifically runners, cyclists and triathletes.[21] The Strava website and mobile app let users track their activities and share and compare their performances with those of other users, called "athletes," in the app. Founded in San Francisco in 2009, the company has a global reach and publishes its app in eleven languages. It is used by both professional and amateur athletes. The app is free but for €59.99 a year Strava also offers a premium membership that gives users additional features.

As a private company, Strava does not publish information about its number of premium users. For this research I created a Strava account. The personal homepage to that account, https://www.strava.com/athletes/29553709, lists me as the 29,553,709th user. It is not clear how many of my predecessors are active users. Strava is not yet profitable and has raised venture capital to compensate its expenses.[22] Strava is one of the last independent fitness trackers. Other trackers were developed or bought up by either technology firms, like Apple (Apple Health), Fitbit (Fitstar), Google (Google Fit), Microsoft (MSN Health & Fitness), Nokia (Nokia Sports Tracker), Samsung (Samsung Health), or sportswear companies like Adidas (Runtastic), Asics (RunKeeper), Nike (Nike Training Club) and Under Armour (Endomondo, MyFitnessPal, MapMyFitness). Following the acquisition of three fitness tracking apps, the American sportswear brand Under Armour issued a press release that states that the apps will help the companies' understanding of the evolving needs of athletes, that is, how they interact, consume and strive to live healthier lifestyles.[23] Kevin Plank, chief executive of Under Armour, said in a 2015 *Financial Times* interview: "We now have the world's largest digital health and fitness community. We believe ultimately this will help us sell more shirts and shoes, reach more athletes and make them better."[24]

In 2013 Strava published its first Global Heatmap showing the aggregated and anonymized activities of its users. The map generated the interest of city planners and departments of transportation, for whom it provided a unique opportunity to see the movements of cyclists through their cities.[25] Measuring movement is a method these planners and departments use regularly. Traditionally the measuring is done by people notating by hand. The dataset of Strava offered a unique opportunity to access many movements, at different times of the day, including routes and speed. In 2014 a new enterprise grew from this interest of cities: Strava Metro uses the data of Strava users to work with cities to improve infrastructure for cyclists and pedestrians.[26] On its website it states that over one hundred transportation planning departments use Strava Metro data. This is a paid service that charges € 0.80 a year for every Strava member being tracked.[27] Keeping the commercial enterprise of Strava Metro in mind, the Strava Global Heatmap is much more than a map that shows the data of its users. And while the introduction pop-up window of the Global Heatmap states that it was made for users to "discover new places to be active,"[28] I cannot but help to see the map first and foremost as a showcase of the technical and commercial possibilities and the data the company has to sell.

Strava not only passively tracked data for Strava Metro, it also actively encouraged users to collect more data. On 10 May 2016, Strava invited cyclists to join the Global Bike to Work Day Challenge. *#CommutesCount*, a short animation that accompanied the campaign, contains the message: "Commuting with Strava makes cities better."[29] If the message had been completely transparent, it would have ended with "and Strava too," given the fact that Strava Metro is commercially exploiting the data it collects. The Strava app gives users the opportunity to not share their data. The app's default setting, however, is to accept the sharing of user data with Strava. Users have to go into the app's settings to disable tracking, but most people do not.

Not all self-tracking practices are taken up voluntarily. Deborah Lupton, professor of sociology at the University of New South Wales, Sydney, identifies five different modes: private, pushed, communal, imposed and exploited self-tracking.[30] In the case of the Strava app, a variety of these types can be observed. Many individual Strava users will use the self-surveillance app to collect information about themselves, raise self-awareness and optimize their lives.

Since Strava is also a social medium, the focus of the tracking goes beyond the individual user. Users are part of a community of trackers. The platform enables the comparing and sharing of data with other members. Another form of communality is the message Strava puts forward in its Commutes Count campaign. In an animation for the 2017 campaign, users are persuaded to take responsibility for their city and take part in an event: "Every time you commute on Strava you create anonymous data. The data show urban planners how to improve your city. Commuting by bike is good for you anyway. Now it's good for your city too. Record your ride on May 11 for Global Bike to Work Day."[31]

The Strava Global Heatmap revealed the locations of secret military bases as the personnel on those bases were recording their activities. Ironically, some military personnel were given the devices to track their activities by the army. In 2013 the US Army provided some of its personnel with fitness trackers to promote a healthier lifestyle.[32] A considerable number of the military personnel can be considered obese.[33] Lupton calls this mode of self-tracking, where the initial incentive doesn't come from the user, pushed self-tracking.[34] Or even imposed self-tracking if the benefit is only to the advantage of others than the user.[35]

Exploited self-tracking are those practices where personal data of users are repurposed for the (commercial) benefit of others. This is the case when Strava Metro sells the behavioral data of users to cities. One could argue that in the anonymized aggregated data it is not possible to find the data patterns of individual users. However, hacking practices and instances where datasets have been combined to uncover secrets show that the boundary between personal small data and anonymous big data is blurry at best, and nonexistent at worst.

Deceiving Familiarity

On 1 November 2017, Strava published an updated version of its Global Heatmap. The map was generated using data of the past two years. In total ten terabytes of raw input, data of 1 billion activities were used to create the map.[36] The map consists of two layers: a background layer containing relief and road information

21 Strava, "About Us."
22 Lassiter III, Sahlman, and Misra. "Strava."
23 "Under Armour Reports Full Year Net Revenues Growth of 32%; Announces Creation of World's Largest Digital Health and Fitness Community."
24 Bradshaw, "Under Armour Snaps up Fitness Apps."
25 Shaw, "The Story behind Strava Metro."
26 "Strava Metro."
27 Olson, "Why Google's Waze Is Trading User Data with Local Governments."
28 "Strava Global Heatmap."
29 "#CommutesCount."
30 Lupton, "Self-Tracking Modes: Reflexive Self-Monitoring and Data Practices," 5.
31 "Commutes Count."
32 Lilley, "20,000 Soldiers Tapped for Army Fitness Program's 2nd Trial."
33 Tilghman, "The US Military Has a Huge Problem with Obesity and It's Only Getting Worse."
34 Lupton, "Self-Tracking Modes: Reflexive Self-Monitoring and Data Practices," 7.
35 Ibid., 9.
36 Robb, "The Global Heatmap, Now 6× Hotter."

Many fitness trackers function in combination with a smartphone. The fitness tracker detects the number of steps taken, distance traveled on foot, number of floors climbed, calories burned, sleep patterns and heart rate. Through a Bluetooth signal, the tracker uses the satellite navigation capabilities of the smartphone. The tracker also uses the phone's connection to the Internet to store and share recorded performances online. In this series of scans, the antennas of a smartphone and fitness tracker are shown. The fitness tracker is the FitBit Flex, the tracker that the US Army gave to parts of its personnel to combat obesity among its troops.

iPhone 5 GPS antenna

from open-source mapping platform Open Street Map and a foreground layer containing the aggregated and anonymized Strava user activities of the past two years. The map is an interactive website that lets users zoom in and out, adjust the transparency and the color of the foreground layer, switch the background layer map for a map or filter the activities that are visible. My reading of the map is based on the default settings of the map, which are also the settings used by Strava in the communication about the map.

I will not dwell on the controversy following military analyst Nathan Ruser's discovery of secret military bases on the Global Heatmap on 27 January 2018.[37] Nor will I discuss the backlash of international news media, the response of the military or the statement and actions undertaken by Strava. Neither will I discuss what the data in the map show or do not show as the majority of Strava users are men, while it also has been argued that the Heatmap only contains activities by affluent runners.[38] For me, the controversy surrounding the discovery of secret military bases is an incident that distracts from the issue at hand: What does the map reveal about Strava and its agenda? And, what to me is a more fundamental issue, can Strava's underlying intentions be observed on the map, and if so, how?

A heat map is a visualization where values are represented by a range of colors. The term was coined and trademarked in 1991 by the then twenty-year-old American student in economics and business science Cormac Kinney for a 2-D visualization of financial market information.[39] This specific type of visualization has an even longer history, dating back to matrix displays, before and after the computer era.[40] The visual format was developed by a number of statisticians, among them the French cartographer Jacques Bertin. I suspect that when Kinney invented the name he was reminded of thermography, the photographic recording of warmth, as the colors of these photographs, ranging from blue for cold via green, yellow and orange to red for warm, correspond with the color palette used in many data visualizations of the type we now call heat maps. The reason for this particular color palette is that it has the largest range of colors, much larger than, for instance, from white via gray to black, and is therefore capable of showing the widest range of values.

In order to make a map of the public activities of Strava users, the data need to be filtered. First a speed threshold was applied to filter out high speeds that probably are the result of a user traveling in a car or plane.[41] Also, nonmoving activities like exercises in a gym are filtered out. Still, the remaining amount of data is so large that it is not possible to generate it all at once. Therefore the map is divided into many smaller square segments—called map tiles—that are rendered separately. This process is repeated for each of the map's twenty zoom levels. At the highest zoom level, the map consists of 30 million tiles.

To ensure that the individual map tiles connect when reassembled as a map, a normalization process is applied in the translation process of data, numbers, to map imagery, pixels. Normalization ensures a uniform distribution of the different visual steps, from dark for low data values to light for high data values. Without the normalization only the single most popular areas on Strava will be rendered. The normalization technique is common in image processing, of photographic images, for instance, where it is called histogram equalization. The effect of the technique is an evenly balanced range of tones and colors that results in a strongly heightened contrast in images that are either very dark or very light. For the individually rendered tiles to connect well, a normalization setting is chosen that filters out activities with little heat. Although the reasoning behind the application of a normalization filter is understandable—without it the map would be very empty in most parts or have areas that do not connect smoothly to neighboring zones—the fact remains that data are manipulated and while the map is quantitatively accurate on a map-tile level, it is not on the level of the map as a whole. Strava admits as much in a blog post and gives two arguments for applying the normalization filter: it ensures a uniform distribution of the color values over the map and Strava "also subjectively finds that it looks really nice."[42]

Another manipulation of data is the introduction of a random offset of a two-meter-wide distribution to all points.[43] Instead of several lines on top of each other for activities on a footpath, for instance, all the points that constitute the lines of the walks are dispersed randomly within two meters of that point. This will make the footpath appear wider and more blurry, it will also result in footpaths that have more activities showing up more clearly on the map as more activities are used to render the line on the map. Another reason for the random offset is that many mobile devices, most notably the iPhone, will occasionally slightly change the GPS signal in urban areas and connect it to road geometry in the device's database rather than the recorded position of the user. As a result, the aggregated activities in those instances are perceived as a thin line along the street. In the Global Heatmap the effect of this error is diminished by adding the random offset.

Historian of data visualizations Michael Friendly has argued that there is a trickle-down effect in the understanding of visualization formats.[44] New types of visualizations are developed by specialists. Over time these improvements of maps and diagrams are implemented in publications, after which it takes some more time for the general audience to grasp them fully. This process takes a considerable amount of time and only today is a general audience familiar with formats developed in the second half of the nineteenth century. According to Friendly, this period is the first golden age of data visualization, an era with many "milestones" in the advancement of data visualization.[45] This golden age was followed by the modern dark ages of data visualization, a period of few new discoveries. Right now, under the influence of digital technologies, we are living in a second golden age according to Friendly, but it will take some time before the improvements and new kinds of visualizations developed today are widely used and even longer for them to be understood by all.

In the Global Heatmap, processes and effects are used that refer to the photographic process. Even the name heat map might refer to photography, as I speculate that it is derived from thermography. The sharply defined background with blurry lines on top suggests a depth of field more closely linked to photography than cartography. The overall image of white blurry lines on a black background is reminiscent of long-exposure photography. The process of image normalization used in the processing of the data is one often used in photographic image editing software. The effect of the image normalization, a stronger contrast, is often used in photography as well as in other graphic reproduction processes. The combined effect of these techniques and effects is an ambiguous image of a photographic map or a cartographic photograph. It is a visualization that we do not know how to read but whose visual language looks familiar.

37 Ruser, "Strava released their global heatmap."
38 Malouff, "Heat maps show where people bike... or at least, where affluent people exercise by bike."
39 US Patent "Heatmaps."
40 Wilkinson and Friendly, "The History of the Cluster Heat Map," 183.
41 Robb, "The Global Heatmap, Now 6× Hotter."
42 Ibid.
43 Ibid.
44 Friendly, "A Brief History of Data Visualization."
45 Ibid., 2.

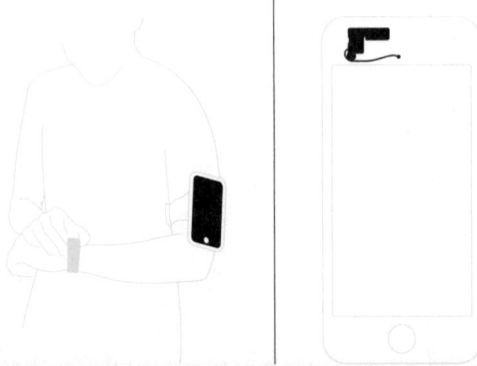

iPhone 5 Wifi and Bluetooth antenna

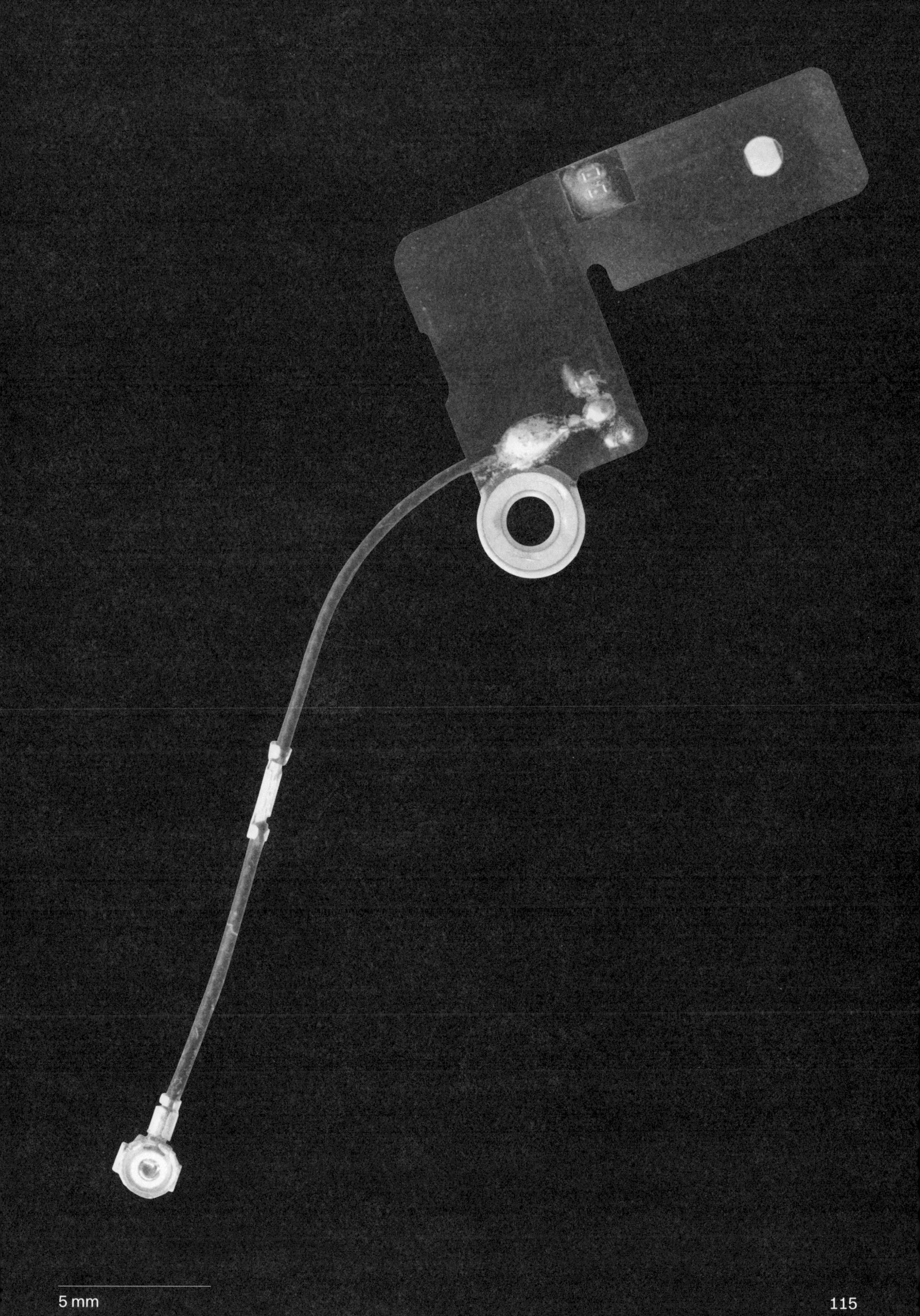

5 mm

Conclusion

In the Strava Global Heatmap there is a disconnection between what you see and what is on the map. The complexity and multilayeredness of the Heatmap become apparent by listing the many discrepancies found in analyzing its full process. There is the difference between the particularity of the user's individual GPS track and the total knowledge pursued by governments and enterprises using the tracks. There is also the opposition between signals of a remote global satellite network and that of an intimate heartbeat that are monitored and combined by fitness trackers and fitness apps. And there is the opposition between nonvisual data and the cartographic visualization that is made from it. As well as the difference of the pretext given and what I suspect to be the true motivation for producing the Strava Global Heatmap: not to show users where they can go, but to show customers where Strava's users are going.

The discrepancy that intrigues me the most in the cartographic rendering by Strava is the difference between the disembodied point of view that the map takes and the content of showing the tracks of bodies moving. American feminist scholar Donna Haraway argues that all vision is embodied.[46] In the 1988 article "Situated Knowledges: The Science Question in Feminism and the Privilege of Partial Perspective," Haraway is critical of the disembodied objective view in science that is also taken in many maps. At the same time, Haraway does not favor the opposite position according to which neutrality does not exist and all is a matter of opinion. Instead, she proposes an approach that is based on "situated knowledges," and urges us to think outside the duality of objectivity and subjectivity. She argues "for a doctrine and practice of objectivity that privileges contestation, deconstruction, passionate construction, webbed connections, and hope for transformation of systems of knowledge and ways of seeing."[47] Haraway contends that by recognizing the impossibility of neutrality and by questioning one's own position, a greater claim to objectivity can be made.

I want to argue that, from a post-representational perspective, the term *situated knowledges* can be used to evaluate the production and use of maps. As Haraway has shown, knowledge is not neutral, it is always situated: the mapmaker always has a subjective position, even when there is no agenda or design strategy, and even if an attempt is made to ignore or disguise it. This is not only true for the producer of the map, but equally true for the user of the map. Like the producer, the user of a map is situated. Users bring experiences and value systems, they access information at a particular time and place, and they too have their particular motives for using a map. The question then remains: If both the production and the use of maps is situated, how then can an exchange of knowledge occur?

The strategies of contestation and deconstruction argued for by Haraway are absent in the Strava Global Heatmap. On the contrary, the map uses a deceivingly familiar formal language. It looks like a map, like a familiar image, but it is a visualization type we are not yet familiar with, the content of which is manipulated, using a visual language with similarities to photographic images that makes it appear well-known. If defamiliarization is the technique of disrupting the user's expectations to stimulate fresh perceptions, than the visual strategies at work in the Global Heatmap do the exact opposite. They render the map in a soothing, familiar form.

On closer inspection, the deceiving familiar strategies are at work on several levels in the case study of the Strava Global Heatmap. For example, the platforms offered by social media are presented as public space, but are in fact private space. They are not town squares but shopping malls that are designed to keep users there, maximizing the number of moments to monetize their presence. In the case of GPS, a defense technology was opened up to empower civilians, but at the same time the move was a shift in power from military to economic dominance. And there are the GPS tracks that are personal expressions of users, selfies of their movements, scratch marks on the maps, that as part of the pursuit for total vision might reveal the patterns of our lives, the places we work, the secrets we keep.

From a post-representational cartography perspective, the deceiving familiarity of the Strava Global Heatmap raises an intriguing question: If a map is seen as a process, as an unfinished product that needs a user to complete it, can an object that is falsely understood by a user to be one kind of map, but is in fact a cartographic representation of something else, still be called a map? To me this is not a binary issue; the question is more interesting than the answer. The same can be said about the other discrepancies I encounter in this research. These are not either-or issues, but matters that are both-and.

I was not expecting it when starting this research into the Global Heatmap, but I found the Strava case study to be an endorsement of the ambiguous strategies I employ in my map design practice, which I presented in the previous chapter. The presentation of complex information asks for a design approach that both presents the content and questions itself, questions the methods and format employed to show it. When there is no clarity, show the opacity. The many layers and intricacies of the Strava case highlight this need. In Donna Haraway's "Situated Knowledges" I found a theoretical foundation for an approach that takes the design of information even one step further: in designing information not only content and form need to be questioned, but the positions of designer and user as well.

At the Visualizing Knowledge 2019 conference at Aalto University Helsinki,[48] Finnish visual artist Antti Tenetz presented his Jalesta/Tracing project in which the movements of wolves and trout in Lapland are visualized.[49] The data derived from GPS collars is presented in multilayered installations consisting of video, photographs and data visualizations. What struck me in his presentation, moved me even, was when he talked about the ethical dilemmas of visualizing the tracks of wolves. Making visible the movements that the animals try to hide, making public where they live for scientific or artistic purposes, also makes the information accessible to hunters. And while I do not want to draw a parallel between Tenetz's project and the Strava Global Heatmap—the Strava users voluntarily share their data and have the right to opt out of any exploitation of their information—Tenetz raises a valuable point. When questioning the contents, methods and formats of visualizations, we should first and foremost ask whether it should be shown at all.

46   Haraway, "Situated Knowledges: The Science Question in Feminism and the Privilege of Partial Perspective," 581.
47   Ibid., 584–585.
48   "Visualizing Knowledge 2019."
49   Tenetz, "Tracing—Jalestaa."

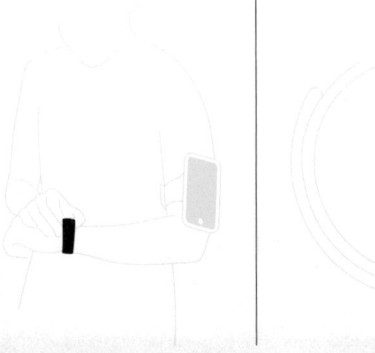

5   The Situation in Syria

In this chapter the cartographic process is considered to comprise two stages. The first is mapping, the surveying of a site and the collecting, selecting and interpreting of data. In the second stage, mapmaking, the data is decoded, edited and transformed into a map. Historically, amateurs—nonspecialists who lack training, knowledge and skills—have always played a role in cartographic processes, but predominantly as surveyors. Recently, however, digital technologies have democratized the tools to create, record, edit, produce and distribute visual information and enabled amateurs to create and publish their own maps. This chapter looks into the practices of amateur conflict mapmakers and their efforts to interpret and transform data into maps. The objective is to find an answer to the question: How do mapmaking practices of specialists and nonspecialists differ in terms of how the work is produced, in terms of the visual strategies that are employed, and in the way the work is made public? And what impact do these possible differences have on how the maps and mapmaking practices are perceived?

This chapter talks about two Thomases, both desk cartographers, one an eighteenth-century Spanish specialist mapmaker who spent most of his mature life working on a map of his home country, the other a twenty-first-century Dutch amateur who published several maps of a conflict in a country three thousand kilometers from his home.

Amateurs and Specialists, Surveying and Mapmaking

"The best way to make a map is by walking and measuring the land, but such a method is not possible for a private individual." This is a quote from Spanish geographer Tomás López (1730–1802).[1] As head of the Gabinete Geográfico, the geography cabinet, López was responsible for the *Atlas geográfico de España*, the first comprehensive and detailed map of Spain, a project started in 1766 and published posthumously in 1804. For reasons of lack of funding, personnel and technical means, rather than having the land surveyed by specialists, López used a "desk cartography" method where an expert cartographer interprets the fieldwork of locals who lack geographic knowledge. In the case of the *Geographic Atlas of Spain*, village priests were approached to answer a questionnaire and draw a map of the territory around their town or village. The survey resulted in hundreds of maps that were inconsistent, as the priests had no scientific training and their level of drawing skills varied. It took López many years to interpret the answers and sketches of the clergymen and incorporate the information in a map.

Ironically, the accuracy of López's atlas of Spain was tested a few years after it was published when the country went to war with France.[2] Napoleon Bonaparte's armies soon found out that the maps lacked precision. The errors were caused by López's nontopographic surveying method. López had learned the method from Frenchman Jean Baptiste Bourguignon d'Anville (1697–1782), one of the eighteenth century's most prestigious cartographers, but had applied it with less rigor, and his instructions to the clergymen who conducted the surveys had not been specific enough.

The example of the *Geographic Atlas of Spain* addresses two dichotomies that are the subject of this chapter. There is the division between mapping and

1 San-Antonio-Gómez, Velilla and Manzano-Agugliaro, "Tomás López's Geographic Atlas of Spain in the Peninsulan War: A Methodology for Determining Errors."
2 Ibid.

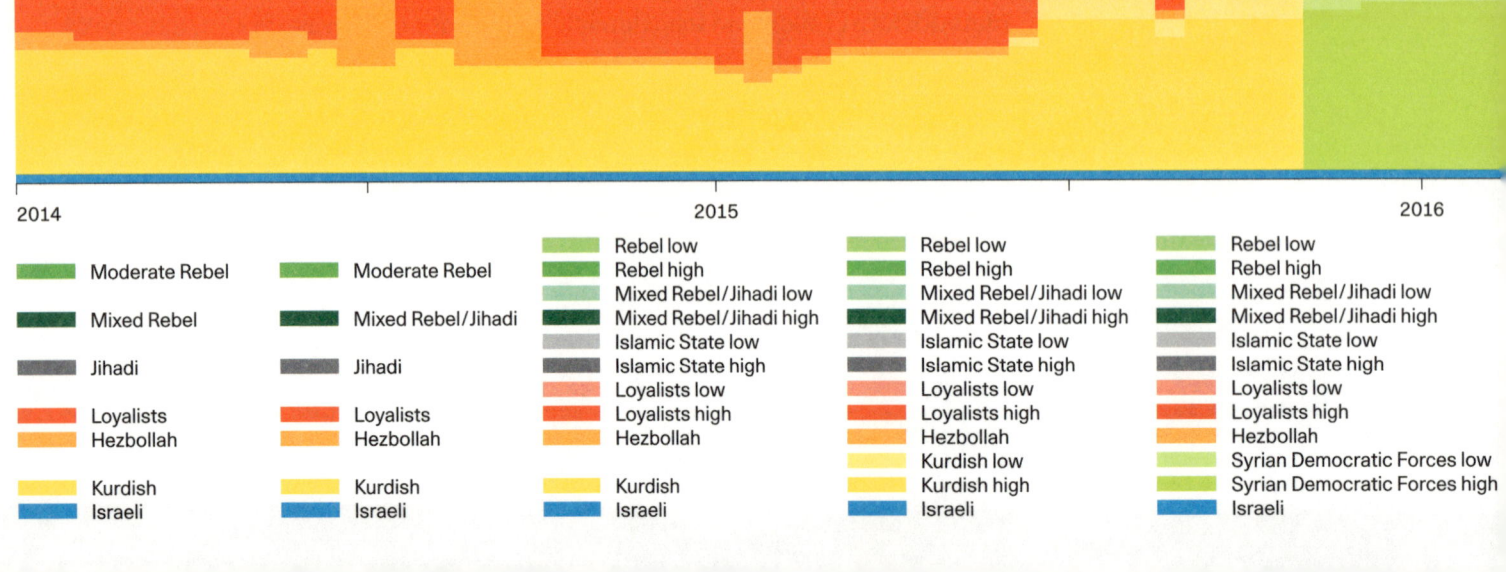

2014   2015   2016

Moderate Rebel	Moderate Rebel	Rebel low	Rebel low	Rebel low
		Rebel high	Rebel high	Rebel high
		Mixed Rebel/Jihadi low	Mixed Rebel/Jihadi low	Mixed Rebel/Jihadi low
Mixed Rebel	Mixed Rebel/Jihadi	Mixed Rebel/Jihadi high	Mixed Rebel/Jihadi high	Mixed Rebel/Jihadi high
		Islamic State low	Islamic State low	Islamic State low
Jihadi	Jihadi	Islamic State high	Islamic State high	Islamic State high
		Loyalists low	Loyalists low	Loyalists low
Loyalists	Loyalists	Loyalists high	Loyalists high	Loyalists high
Hezbollah	Hezbollah	Hezbollah	Hezbollah	Hezbollah
		Kurdish low	Kurdish low	Syrian Democratic Forces low
Kurdish	Kurdish	Kurdish high	Kurdish high	Syrian Democratic Forces high
Israeli	Israeli	Israeli	Israeli	Israeli

2017   2018

This is a timeline of the 64 "Situation in Syria" maps that Thomas van Linge published on Twitter between January 2014 and April 2018. In each map, the area a certain faction controlled is measured. This is represented by a color code for which the colors of the "Situation in Syria" maps have been used. The horizontal width of each color bar corresponds to the time between the date of publication and the moment a new edition of the map was published. The equal widths of the color bars show the great regularity of Van Linge's updates. The legend on the left page is based on the terminology that Van Linge uses in the maps.

mapmaking: the difference between the collecting of data to describe a situation and on the other hand the editing and transforming of that data into a map. The other is the distinction between specialists and nonspecialists, between those who are trained, have knowledge and skills and those who lack these but who instead have knowledge of, or access to, a site, or are willing to make inexhaustible efforts to collect and compare data and convert it into a map. The Spanish example also addresses the role a map plays in claims of statehood. Most maps in this chapter deal with this issue, but this will not be a topic of investigation.

The method of drawing a map without surveying a site, but instead by interpreting several data sources, has been called armchair cartography, remote cartography and desk cartography. None of these names is entirely satisfactory to me. The adjective *armchair* is often used to indicate that someone has little or no practical knowledge or experience but still regards herself to be a specialist. This is not the case with Tomás López, who had studied with one of the most renowned cartographers of his time and deliberately chose to not go out to survey the whole of Spain for practical and economic reasons. As most mapping nowadays involves satellites, either as recorders or transmitters of data, the name *remote mapping* does not seem precise enough to describe the phenomenon of a cartographic method without on-site surveying. Desk cartography seems the most accurate description. However, I prefer the term *desktop cartography*, which refers both to a piece of furniture and the working area of a computer screen as well as to desktop publishing (or DTP), the creation of documents using page-layout software on a computer. Most cartography nowadays is desktop cartography. With aerial photography and satellite-sensing technologies it has become less important to survey on-site. Evolving technologies also resulted in the emergence of on-site mapping practices, such as fitness tracking, which was the topic of a previous chapter.

In this text, which addresses the dichotomies of mapping/mapmaking and specialists/nonspecialists, I will look at a specific case of desktop cartography: amateur conflict mapmakers who map sites that are dangerous to access, about which opposing territorial claims are made, and that are in constant flux. First I will address the phenomenon of conflict mapmaking, comparing specialist and nonspecialist practices that map situations of political, social and/or geographic conflict. Then I will delve into how these maps are made public, shared and become part of a public debate. I will examine the visual strategies of the cartographic language of amateur conflict mapmakers. In a second section I will discuss more in detail the practice of one specific amateur conflict mapmaker.

Mapping Conflicts

In a 2017 article, Dietmar Offenhuber, head of the Information Design and Visualization program of Northeastern University in Boston, who has a background in urban planning and whose research focuses on the relationship between design, technology and urban governance, addresses the challenges, including ethical issues, of cartographic representations of the self-proclaimed Islamic State (IS) in Iraq and Syria.[3] Offenhuber argues that the sovereignty of the IS territory is symbolically challenged through cartographic choices that reflect the diverse interests of the mapmakers. Maps of the IS territory made by Western news organizations depict densely populated cities and the transport routes connecting them, but not the deserts or sparsely populated land in between. The mesh-like structure makes the territory look unstable, fluid and ambiguous.[4] The open structure visually depreciates the IS territory and thus avoids similarity to a traditional state map.[5] Offenhuber compares the cartography of news organizations with the maps that IS itself produces. In these maps the "caliphate" is depicted as a closed and contiguous shape, a unified state covering large parts of Syria and Iraq, even including sixteen provinces, each with its own name.[6] Putting a name on a map is the equivalent of planting a flag on a piece of land, it is making a claim. Offenhuber presents a third map that, like the IS map, displays the area of the "caliphate" as large as possible, although not for ideological but for economic reasons. The Coalition for a Democratic Syria, a Syrian-American organization advocating for expanded US support of the Syrian opposition, uses for its map the provinces as the smallest cartographic unit.[7] The area controlled by IS therefore looks impressively large. The map might be persuasive to get support for the purposes of its mapmaker; the territory, however, consists for a large part of empty desert. These three examples show that a map always represents the ideological, economic or other concerns of the mapmaker.

In addition to the practices listed above, Offenhuber's article examines maps made by amateur conflict mapmakers[8] and visual forensic experts, who use and cross-reference data from social media such as movies and photographs taken by mobile phones and drones, georeferenced twitter messages and satellite imagery. The aggregated data is verified, timestamped, geotagged and used as the basis for the cartographic and other visualizations that these practices produce. Compared with the official conflict maps of news organizations, the work of the amateur conflict mapmakers often looks more raw and unprocessed.[9] This absence of refinement might originate in a lack of training in graphic design or cartography, Offenhuber argues, however, that the visual strategies developed by the nonspecialists follow a visual logic that serves the purpose of presenting evidence. The visual vocabulary of the amateur conflict mapmakers focuses on showing the employed methods of cross-referencing original footage. If a user is in doubt, she can utilize Google Earth to look up the location and compare it with the map and its highlighted features. Offenhuber states that this strategy fits the endeavors of amateur conflict mapmakers to challenge official reports, expose attempts to mislead and to identify misinformation.[10]

One of the practices to which Offenhuber refers is the investigative journalist's website Bellingcat. Today a network of staff and contributors in more than twenty countries, the platform was founded in July 2014 by British citizen journalist Eliot Higgins. I will discuss a blog post on the Bellingcat website by Higgins from 15 July 2014, the early days of the platform, about a chemical weapon attack in Ghouta, Syria.[11] The post consists of a text interspersed with annotated satellite images and video stills. At one point in the blog post a zoomed-in fragment of a satellite image is introduced as a "piece of the puzzle" and that is how the post feels: image by image, the reader is led along one piece of evidence to the next until the inevitable conclusion.[12] The text reads as a voice-over to a slideshow of annotated images. The imagery is crude: blurry, different in size, occasionally consisting of

3 Offenhuber, "Maps of Daesh: The Cartographic Warfare Surrounding Insurgent Statehood."
4 Ibid., 2.
5 Ibid., 4.
6 Ibid., 6.
7 Ibid., 12.
8 Instead of the term *mapmaker* Offenhuber uses the term *mapper* to describe the amateurs who map conflicts. While the Oxford English Dictionary gives two different definitions of the verb "to map": to represent an area on a map and to record in detail the spatial distribution of something, the OED only gives one definition for the noun "mapmaker": a person who draws or produces maps, only covering the first definition of the verb. This goes back to the opening line of this chapter in which I state that the cartographic process consists of two stages, the surveying of a site and the collecting, selecting and interpreting of data which I name *mapping* and the decoding, editing and transforming of data into a map, which I name *mapmaking*. The term *mapmaker* seems more focused on the second stage, on the editing and transforming of data, and seems to neglect the collecting and surveying. To me it seems that Offenhuber by using the term *mapper* is putting more emphasis on that first stage of the cartographic process. In this chapter, however, I will use the term *mapmaker* to avoid the impression that the maps of amateur conflict mappers, as Offenhuber calls them, are less constructed, more unmediated than the maps of traditional mapmakers.
9 Offenhuber, "Maps of Daesh: The Cartographic Warfare Surrounding Insurgent Statehood."
10 Ibid., 19.
11 Higgins, "Identifying Government Positions during the August 21st Sarin Attacks."
12 Ibid.

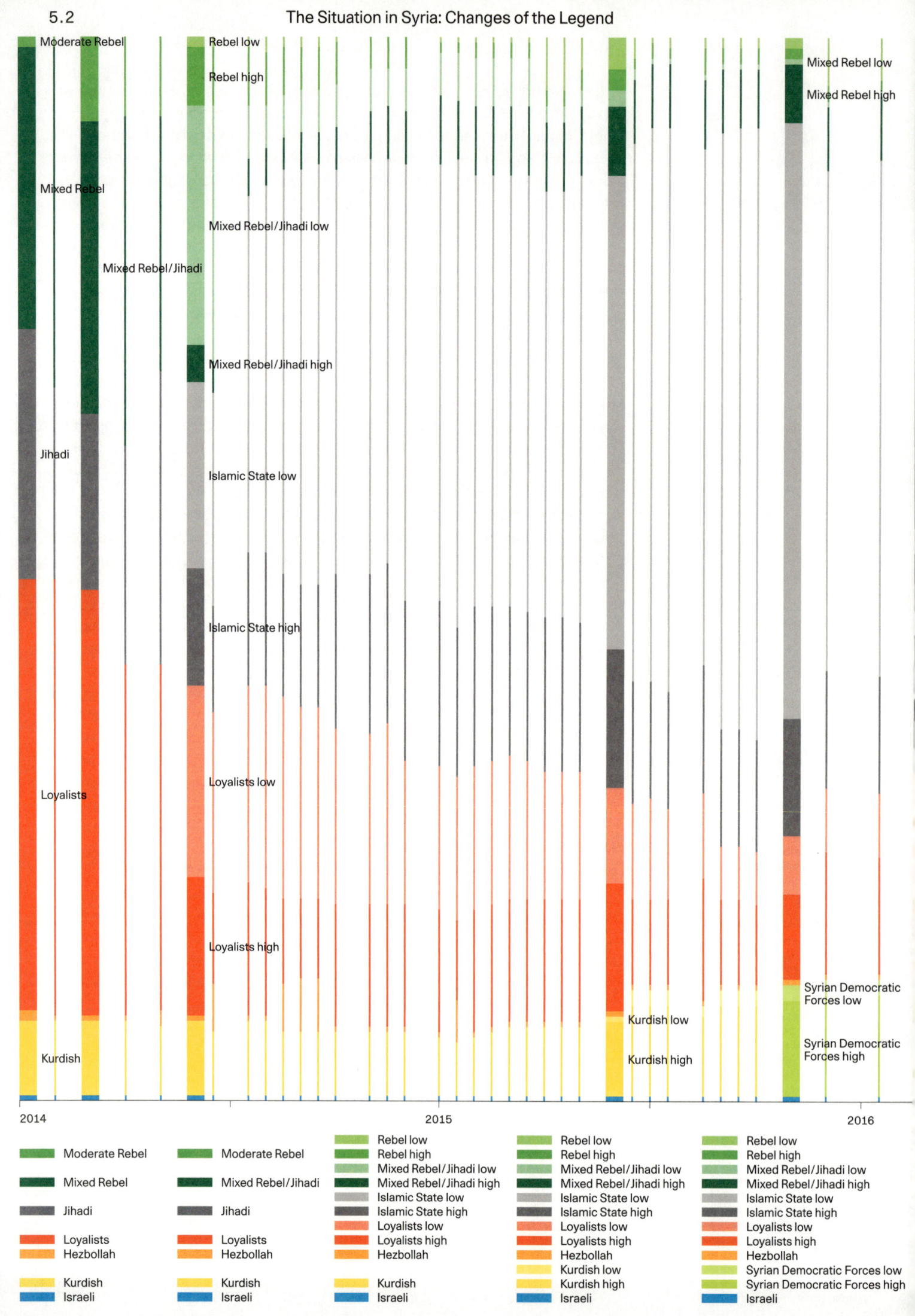

5.2

The Situation in Syria: Changes of the Legend

Moderate Rebel

Mixed Rebel

Mixed Rebel/Jihadi

Rebel low

Rebel high

Mixed Rebel/Jihadi low

Mixed Rebel/Jihadi high

Islamic State low

Islamic State high

Jihadi

Loyalists

Loyalists low

Loyalists high

Kurdish

Kurdish low

Kurdish high

Mixed Rebel low

Mixed Rebel high

Syrian Democratic Forces low

Syrian Democratic Forces high

2014   2015   2016

Moderate Rebel	Moderate Rebel	Rebel low	Rebel low	Rebel low
Mixed Rebel	Mixed Rebel/Jihadi	Rebel high	Rebel high	Rebel high
		Mixed Rebel/Jihadi low	Mixed Rebel/Jihadi low	Mixed Rebel/Jihadi low
		Mixed Rebel/Jihadi high	Mixed Rebel/Jihadi high	Mixed Rebel/Jihadi high
Jihadi	Jihadi	Islamic State low	Islamic State low	Islamic State low
		Islamic State high	Islamic State high	Islamic State high
		Loyalists low	Loyalists low	Loyalists low
Loyalists	Loyalists	Loyalists high	Loyalists high	Loyalists high
Hezbollah	Hezbollah	Hezbollah	Hezbollah	Hezbollah
Kurdish	Kurdish	Kurdish	Kurdish low	Syrian Democratic Forces low
			Kurdish high	Syrian Democratic Forces high
Israeli	Israeli	Israeli	Israeli	Israeli

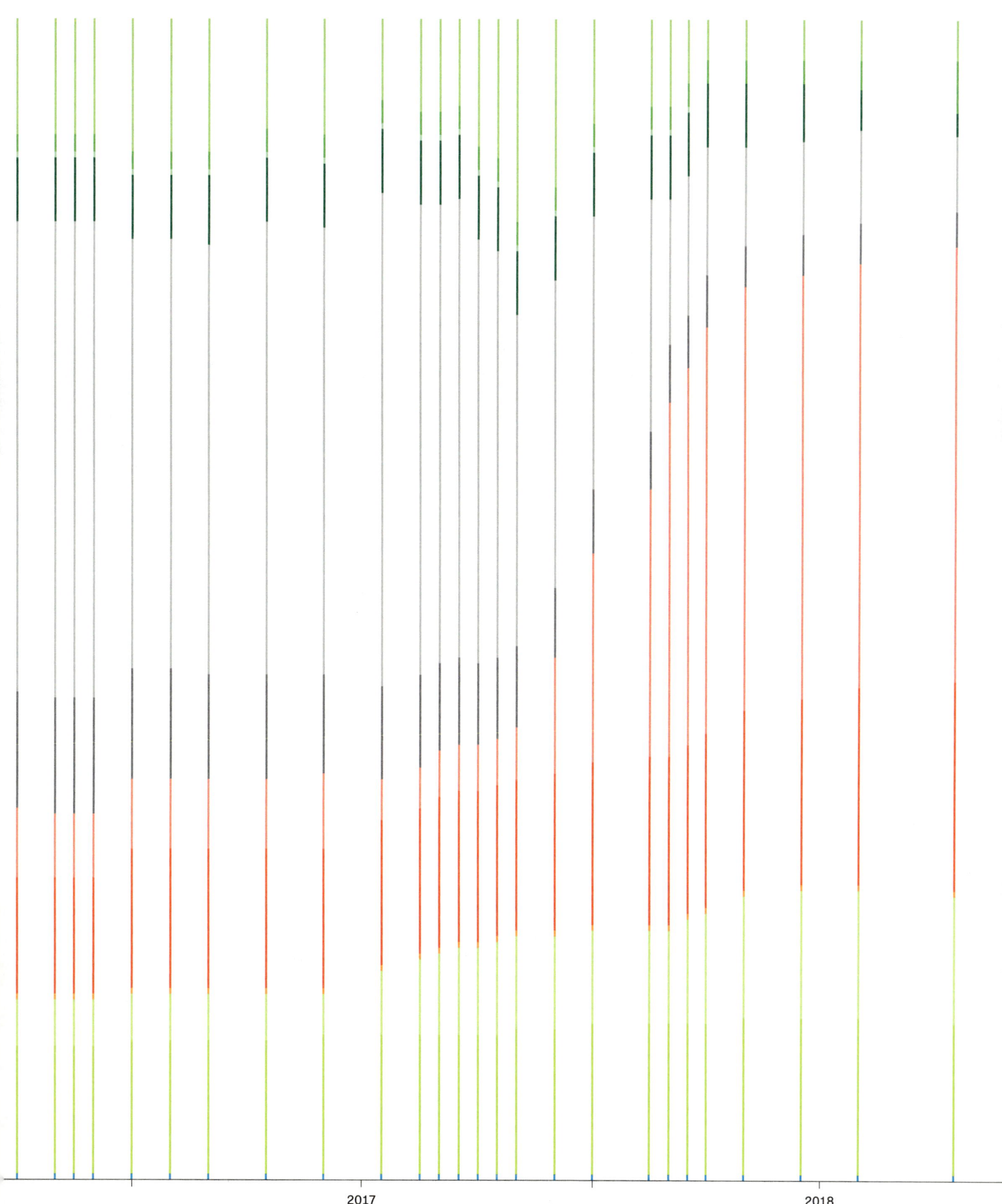

2017   2018

Thomas van Linge adapted the legend of the "Situation in Syria" maps several
times. This timeline shows these changes. In June 2014, Van Linge introduced
a second lighter shade of each legend unit to distinguish between areas that
are densely populated and areas with low population density. Other changes
relate to the naming of the legend.

127

juxtaposed fragments. The annotations are mostly symbols with texts referring to dates, camera positions and names of sites. There are also overlays of transparent-colored shapes marking areas. Circles of various diameters demarcate the site of the attack. The annotated images look unembellished and matter-of-fact, not made to be dwelled upon but to serve a purpose in constructing a narrative in which the maps are used as arguments.

According to Offenhuber, the visual strategies of amateur conflict mapmakers emphasize post-representational aspects of cartography. The maps they produce are not intended to be universal representations, but rather play an ephemeral and circumstantial role in a certain specific public debate or context.[13] The output of these practices is used for exchanges on online platforms like Twitter, they are constantly adjusted and updated and are never considered to be finished. They are therefore by definition processual. Maps produced by official news sources are updated just as often, but the difference between these and amateur conflict maps is that the maps of the second category are embedded in a public exchange that is easily accessible for all and without hierarchical differences in the exchange, as opposed to the dialogue between an individual and a media organization of reputation.

Referencing the semiotic theory of American philosopher Charles Sanders Peirce (1839–1914), Offenhuber indicates what to him is the difference between traditional cartography and maps of amateur conflict mapmakers. Traditional maps use symbolic and iconic signs based on conventions and signs that resemble the object of reference. On the other hand, conflict maps—assemblages of annotated photographs, satellite imagery and video stills—point out what is visible, draw connections and show relations. Whereas traditional maps are symbolic abstractions, conflict maps are "indexical visualizations," according to Offenhuber. *Indexical* is a term of Peircean semiotics to indicate that a sign points to a phenomenon and emphasizes a causal relationship.[14] Offenhuber gives the example of a white dot on a satellite image that "is a consequence of light reflected from an object and registered by the satellite's optical instrument."[15]

Although I follow Offenhuber's analysis that there is a difference between traditional and conflict maps, I find the term *indexical visualizations* for the latter problematic, as it falsely suggests these are unfiltered, unconstructed images. Seeing is not neutral. In his book *Representing and Intervening*, Canadian philosopher of science Ian Hacking argues that images produced by optical instruments like binoculars and microscopes are constructed.[16] Hacking writes extensively about the interventions needed to see with microscopes.[17] Lenses, stains to highlight certain parts of cells (some of these being so toxic that they destroy the tissue), and the flattening of material between glass slides are examples of interventions in microscopic observation. If we take Offenhuber's satellite image example, then there, too, optical instruments intervene in the process of seeing. Lenses create distortions of colors and shapes. Also, the angle of the camera and the relief of the Earth's surface cause misrepresentations. Satellite images are often "orthorectified," that is, they are corrected for topographic relief, lens distortion and camera tilt, before they are used as "map accurate" background images in the production of maps.

American philosopher and pedagogical theorist John Dewey (1859–1952) believes that many epistemologies are based on the mistaken analogy between knowing and seeing an object. Knowing is conceived as what is "supposed to take place in the act of vision."[18] Dewey is critical of the model of knowing as a passive relation between the knower and the object known, which he calls the "Spectator Theory of Knowledge."[19] Peirce, Dewey and also Hacking are representatives of the philosophical tradition called pragmatism that understands knowing the world as inseparable from agency in it.[20]

To me, the difference between traditional cartography and conflict maps lies not so much in the visual strategies of the amateur mapmakers, but in the fact that the maps are embedded as arguments in a public debate. Not how it looks, but how it is shared and distributed enables the work of amateur mapmakers to be part of an ongoing dialogue in which the claims made on (or by) a map can be challenged. This becomes apparent when we look at a practice that uses similar visual strategies as the amateur conflict mapmakers, but uses different platforms and contexts to make its work public.

Forensic Architecture is the name of a London-based research agency that uses architectural methods like digital and physical models, 3-D animations, and cartography to investigate human rights or environmental cases that are not adequately addressed by the state in which they took place.[21] The visualizations[22] produced by Forensic Architecture are montages of diverse overlapping materials, footage from mobile phones, drones, security cameras and satellites, 3-D computer models and annotational layers. There are similarities with the assemblages of amateur conflict mapmakers. However, the work of Forensic Architecture is more skilfully produced and has a higher degree of sophistication in terms of how it communicates its message. I am a critical fan of the work of Forensic Architecture; I am fascinated by the methods it employs, because the use of actual visual material gives its output a sense of authenticity, immediacy and urgency. The blurry recordings from security cameras that provide the raw material for its work feel unstaged and the issues investigated are matters of life and death. They are presented in a dry, matter-of-fact way, so that the conclusions seem inevitable. However, I have doubts about the unequivocalness of the conclusions of the investigations of Forensic Architecture and, especially, the supposed unambiguity of the design of its presentations.

In January 2019 I saw Forensic Architecture's exhibition *Forensic Justice* at "base for art, theory, and social action" BAK in Utrecht.[23] It was not the first time I had seen an exhibition with works of Forensic Architecture, but earlier shows included a single work by the agency. The Utrecht exhibition contained eight investigations. Seeing them together made me shift my focus from the individual investigations to the overall approach, to the methodology, the visual strategies and the formats. Forensic Architecture avoids a term like *project* on its website or in its monograph.[24] This word might suggest that a particular work is finished. Instead, it uses *investigation* to describe its output.

One of the investigations presented at the exhibition in Utrecht was a video titled "Pro-Government Strikes on M2 Hospital—Aleppo, 2016," which deals with the attacks on the Omar Bin Abdul Aziz Hospital, also known as M2, in Aleppo, Syria, between June and December 2016.[25] The video begins with a voice-over

13 Offenhuber, "Maps of Daesh: The Cartographic Warfare Surrounding Insurgent Statehood."

14 Buchler, *Philosophical Writings of Peirce*, 107.

15 Offenhuber, "Maps of Daesh: The Cartographic Warfare Surrounding Insurgent Statehood."

16 Hacking, *Representing and Intervening: Introductory Topics in the Philosophy of Natural Science*.

17 Ibid., 186–209.

18 Dewey, *The Quest for Certainty: A Study of the Relation of Knowledge and Action*, 26.

19 Ibid.

20 *Stanford Encyclopaedia of Philosophy*, "pragmatism," accessed 6 September 2019, https://plato.stanford.edu/entries/pragmatism.

21 From the "about" page of the Forensic Architecture website: "Forensic Architecture (FA) is a research agency, based at Goldsmiths, University of London. We undertake advanced spatial and media investigations into cases of human rights violations, with and on behalf of communities affected by political violence, human rights organizations, international prosecutors, environmental justice groups, and media organizations." Forensic Architecture, "About."

22 In this chapter I will use *visualization* as specified in the Oxford English Dictionary's first definition of the term, a "representation of an object, situation, or set of information as a chart or other image," and not as the OED's second description, "the formation of a mental image of something." For the verb *to visualize* I use the second definition in OED, to "make something visible to the eye," rather than the OED's first description, to "form a mental image of."

23 BAK, "Forensic Justice."

24 Weizman, *Forensic Architecture: Violence at the Threshold of Detectability*.

25 Forensic Architecture, "Airstrikes on M2 Hospital."

2014   2015   2016

Retweets
— Likes
/ Replies

Islamic State low
Islamic State high

This overview shows the response on Twitter to the "Situation in Syria" maps compared with the amount of area that the jihadist group Islamic State controlled. This timeline shows a parallel between the rise of the Islamic State and the Twitter-likes and retweets of the "Situation in Syria" maps.

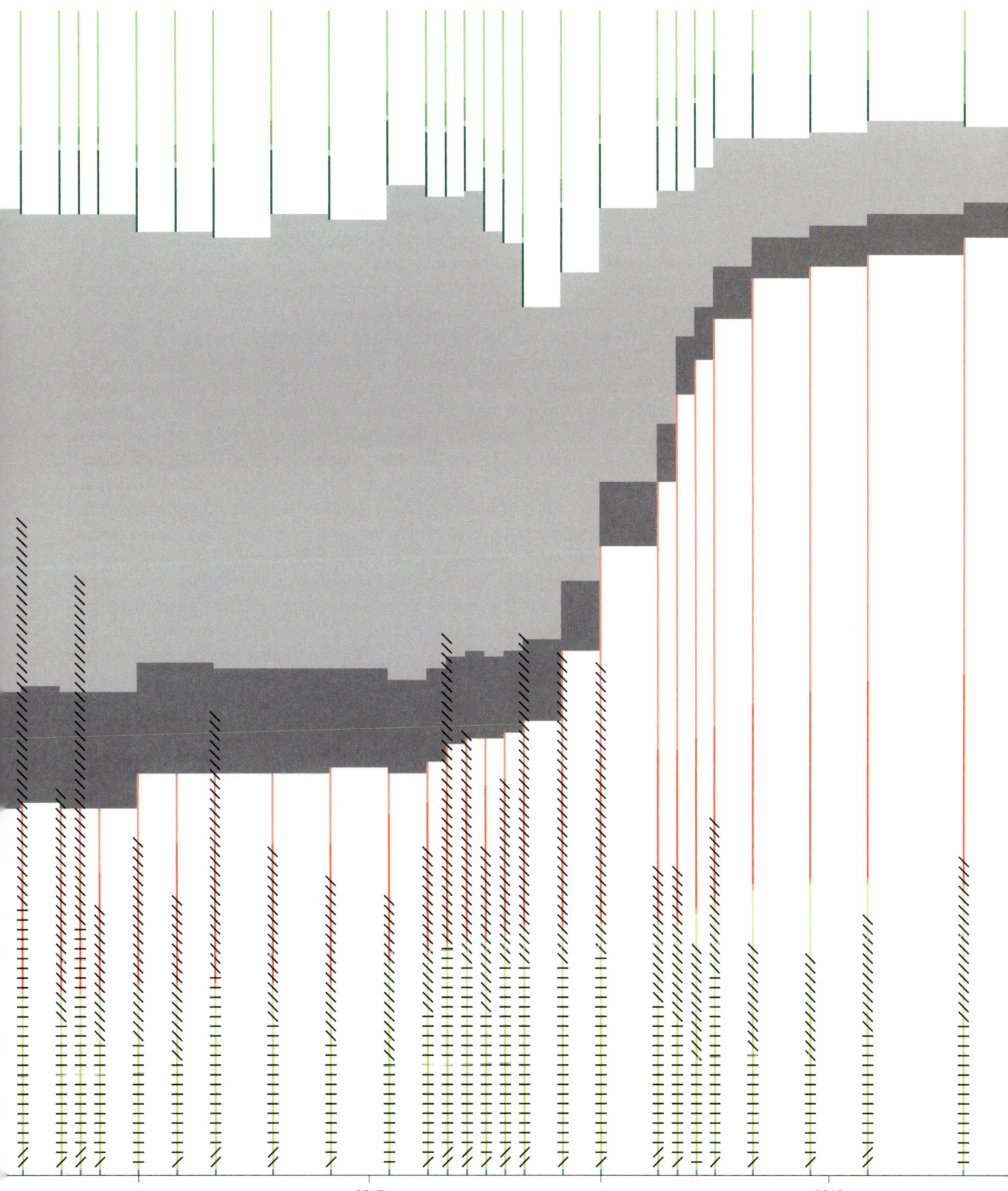

2017

2018

stating that according to the Syrian American Medical Society, the M2 Hospital was subject to fourteen strikes by pro-government forces in 2016. No further proof for this claim is given. The video shows an animation of an architectural computer model of the M2 hospital and its surroundings, reconstructed using satellite photographs, CCTV and handheld camera footage. The computer rendering is light gray and partly filled in with still images from the videos. The animation navigates between the computer model and video and photo footage that is placed inside the model. The animation is a continuous shot, starting inside the hospital at the moment the 16 July 2016 attack took place, moving through the hospital, leaving the building and going into the street, looking back at the building, moving up, looking down and zooming out showing a satellite photo of the city of Aleppo. During the whole video, the animation is accompanied by a voice-over explaining the methods of making the model and about the events shown. It is striking that the video offers no room for doubt. Claims are made without rebuttal. Also what I would call the design of the video—the cool light-gray computer model, the continuous movement of the camera, the compactness of the narrative—communicates self-assurance and control and contributes to this air of doubtlessness.

The inquiries exhibited in Utrecht looked more or less similar in terms of the amount of data collected, the speculations on what took place at the situation that was investigated and the degree of elaboration of the presentation design. Seeing several projects exhibited together, I could not help but think that these investigations were completed. Nothing in the Utrecht exhibition suggested that the research was still going on or that there was additional material that needed to be processed. In the way the inquiries were shown in the exhibition, doubt was only raised about the official accounts of the cases that were researched, not about Forensic Architecture's own methods, the extent of its knowledge about the research subject, or the format in which the findings are shown. This might be caused by the presentation of the projects. The absence of doubt was felt to a far lesser degree when I viewed the same projects on the Forensic Architecture website. In the online context, a user has access to the employed media and resources, can find information about commissioners and funding behind the project, and is able, through e-mail and social media, to get in touch with the investigators and to share the information.

In previous chapters, I used the terms *ontological* and *ontogenic* to describe the difference between critical cartography and post-representational cartography. *Ontological* refers to how things are, *ontogenic* to how things become. My critique of the work of Forensic Architecture and of the work of many amateur conflict mapmakers is that they approach the map as ontologically secure: I fail to see any doubts about the format used and about its status as being complete. Many critical cartographers question the event they are investigating, but not the format they employ. They suggest that a map is a map and it is constant. In that sense their work has a positivist approach that is similar to representational cartography, and uses a comparable but updated technicist language. In the visual language of the maps a variety of heterogeneous image recordings are layered. It is a forensic style equivalent to the use of a handheld camera in film to convey a sense of reality. Only when embedded in a particular context, and when the provenance of the data and the motives of the mapmaker are known, can a map be held accountable for its content.

Offenhuber calls the work of amateur conflict mapmakers and forensic specialists "presentations" rather than "representations."[26] He uses the term *presentation* for the work of amateur conflict mapmakers because their visualizations "point to what is already visible," unlike traditional cartography's use of symbols and abstractions to *represent* a situation or an area of land.[27] I find the distinction between *presentation* and *representation* a problematic one in this regard. The term *presentation* suggests here that conflict maps are unmediated and unconstructed. I disagree: I regard the use of visual material of amateur conflict mapmakers and forensic specialists as constructed, not as authentic, unmediated or objective, and therefore view their output as representations.

In my view, the key difference between the visualizations of amateur conflict mapmakers and forensic specialists on the one hand, and more conventional cartography on the other is the emphasis put on visual evidence by the former. In the two examples I gave earlier, Bellingcat's reporting on the 2014 chemical attack in Ghouta, and Forensic Architecture's video of the investigation in the 2016 M2 hospital attacks, visual proof is extensively discussed. In fact, the majority of the reporting consists of a detailed presentation of the evidence, while significantly less attention is paid to providing an overview or answers. In a lecture at the TU Berlin in 2018, author, cultural scientist and curator Anne Huffschmid talked about her concept of "visibilization," making visible what is hidden, as opposed to visualization, representing information.[28] Huffschmid used this term in relation to her work on *desaparecidos*, the missing, in Mexico, people who as a result of the current escalation of violence in the country have been violently kidnapped, killed and have disappeared. The purpose of her work is the retrieval of anonymous dead people into the social space and the visualization of the crimes to which they have fallen victim.[29] The term *visibilization* is appealing because of its emphasis on the process of unearthing data and on visualization as evidence. To me, *visibilization* is an appropriate term to describe the work of both the amateur conflict mapmakers and forensic specialists. Rather than Offenhuber's distinction between representation and presentation, I propose the dichotomy *visualization* and *visibilization* to describe the difference between traditional cartography, as used in news media, by political institutes and the state, and the maps produced by amateur conflict mapmakers and forensic experts like Forensic Architecture.

One issue in Offenhuber's text that deserves further exploration is the "raw and unprocessed look of the work of amateur conflict mapmakers."[30] Offenhuber speculates that the absence of refinement originates in a lack of training in graphic design or cartography, but also notes that notwithstanding the raw appearance of the maps, their design follows a visual logic that serves the purpose of presenting evidence. In the next section I will look at the processes and tools used in one specific practice of an amateur conflict mapmaker and how these inform the design of the maps.

The Situation in Syria

"IMPORTANT: map about the current situation in #syria. green = regime, brown = #FSA, blue = contested" is the text that accompanied the first map of Syria

26 Offenhuber, "Maps of Daesh: The Cartographic Warfare Surrounding Insurgent Statehood."
27 Ibid.
28 Huffschmid, "Reconstructing Conflict: Mapping as Materialization of Contested Memories and Invisibilized Crime."
29 Huffschmid and Braig, "'Knochenlesen' als Grenzüberschreitung: Forensische Anthropologie als Beitrag zur Gewaltverarbeitung und transnationaler Wissenstransfer, am Beispiel des argentinischen EAAF (Mexiko, Spanien)."
30 Offenhuber, "Maps of Daesh: The Cartographic Warfare Surrounding Insurgent Statehood."

× Koens, "Deze
× "Nederlandse kaa
× "Thomas van Linge"
× "Le dessous des cartes e
× "Les cartes du conflit syri
× Ley, "Kartograf des Krieg
× Février, "A 19 ans, il est le
× Huchon, "Carte jeune pou
× Ricciardelli, "This teenage
× van Huët, "Thomas (19) zi
× Delille, "Le meilleur cartograp
× "Le meilleur cartographe de
× Kuntz, "The Dutch Teen Who
× De Weerd & van Houtum, "Waarom
× Egberts, "18-jarige Amsterdammer br
× "18-jarige kaartenmaker verovert de v
× Westcott, "The High School Student Who

2014   2015   2016

In this timeline articles and news items about Thomas van Linge are compared with the amount of area that the jihadist group Islamic State controlled. This overview shows a parallel between the rise of the Islamic State and the media attention for Thomas van Linge and his "Situation in Syria" maps. The parallel can also be seen in the news items themselves. In the articles a contrast is created between the violence of war and the territorial ambitions of Islamic State and a teenager, living with his parents 3,000 km from Syria, who makes maps from his bedroom using simple means.

⬜ Islamic State low
⬛ Islamic State high

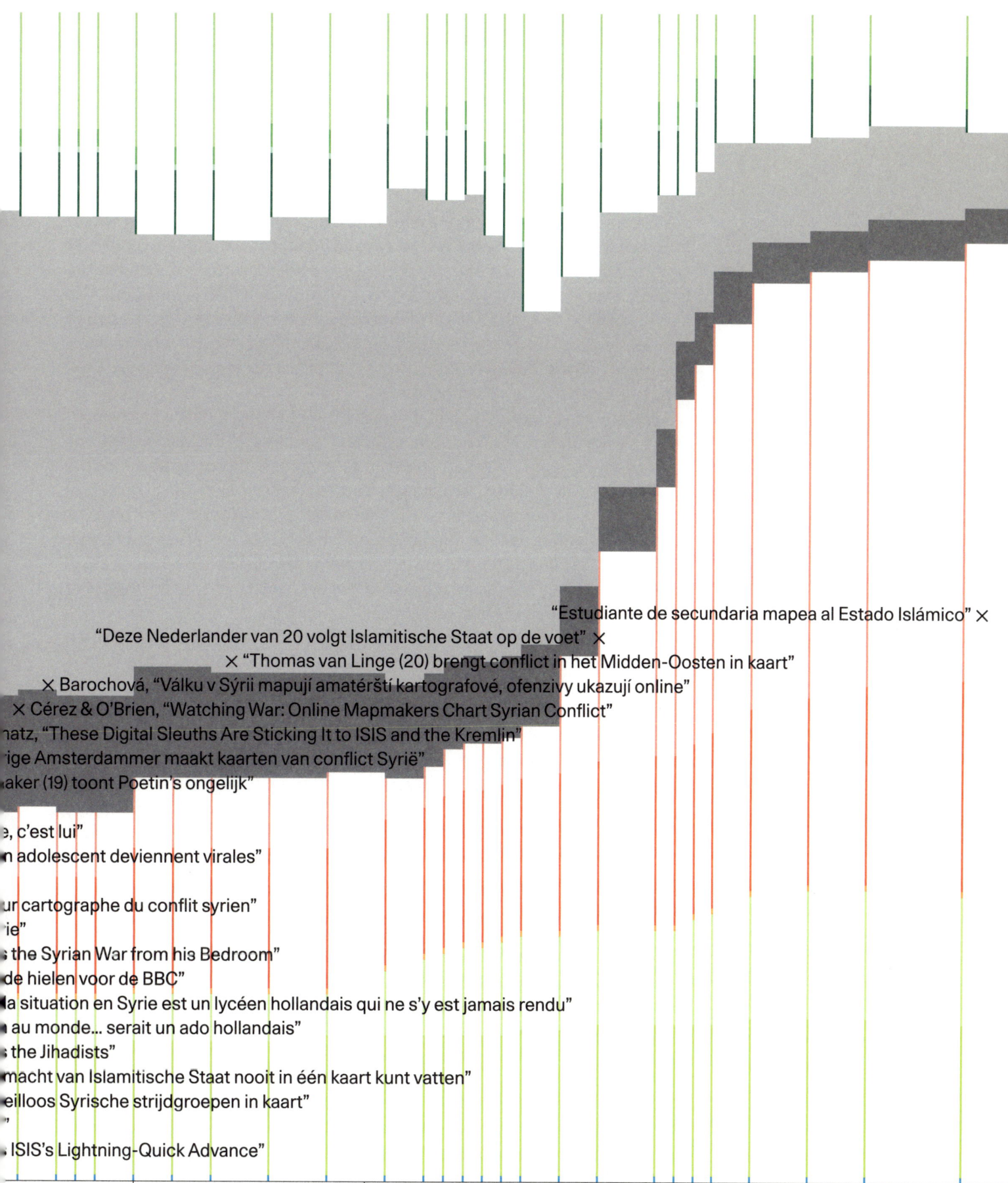

"Estudiante de secundaria mapea al Estado Islámico" ✕

"Deze Nederlander van 20 volgt Islamitische Staat op de voet" ✕

✕ "Thomas van Linge (20) brengt conflict in het Midden-Oosten in kaart"

✕ Barochová, "Válku v Sýrii mapují amatérští kartografové, ofenzivy ukazují online"

✕ Cérez & O'Brien, "Watching War: Online Mapmakers Chart Syrian Conflict"

natz, "These Digital Sleuths Are Sticking It to ISIS and the Kremlin"

rige Amsterdammer maakt kaarten van conflict Syrië"

aker (19) toont Poetin's ongelijk"

e, c'est lui"

n adolescent deviennent virales"

ur cartographe du conflit syrien"

ie"

s the Syrian War from his Bedroom"

de hielen voor de BBC"

la situation en Syrie est un lycéen hollandais qui ne s'y est jamais rendu"

au monde... serait un ado hollandais"

s the Jihadists"

macht van Islamitische Staat nooit in één kaart kunt vatten"

eilloos Syrische strijdgroepen in kaart"

"

ISIS's Lightning-Quick Advance"

2017   2018

posted on Twitter by Dutch political science student Thomas van Linge on 24 June 2013.[31] Van Linge's map documents the situation in Syria where, since the Arab Spring in 2011, an ongoing civil war is being fought between the Syrian Arab Republic, an alliance of opposition rebel groups including the Free Syrian Army, jihadist groups, mixed Kurdish-Arab Syrian Democratic Forces and ISIS, the Islamic State in Iraq and Syria. Iran, Russia, Turkey, the United States and other countries are involved or provide support to one of the factions. The *Encyclopaedia Britannica* lists the Syrian Civil War as the second deadliest war of the twenty-first century, estimating that one in ten Syrians have been killed or wounded by the fighting, causing at least 470,000 deaths.[32] After the first map in June 2013, Van Linge published several dozens in the subsequent years. During the same period he also posted on his Twitter feed maps of Iraq, Ukraine, Libya and other regions. The maps of Syria, however, are his most important subject. For this text I will concentrate on this particular series.

Thomas van Linge was a sixteen-year-old high school student when he published his first map in 2013. Two years earlier his interest in the Middle East was sparked after seeing a TV report on the brutal suppression of protests during the 2011 Egyptian Revolution.[33] Van Linge started following the events in Egypt, Libya and Syria during the so-called Arab Spring, the series of antigovernment protests across North Africa and the Middle East of the early 2010s.[34] Frustrated about not being able to find a map of Syria that makes a clear distinction between areas controlled by moderate Syrian rebels and that by jihadist movements like Jabhat al-Nusra, he decided to make one himself.[35]

Van Linge uses a variety of sources for his maps: social media like Twitter, Facebook and YouTube, local news media, the official accounts of the various forces as well as personal contacts with activists in the Free Syrian Army in Aleppo and in the Kurdish region north of Aleppo.[36,37] Van Linge estimated he had around one thousand sources for his maps to confirm claims of territorial control. Some of the sources he found are retweeted to Van Linge's Twitter account. Since joining Twitter in 2013, Van Linge has sent an average of twenty tweets and retweets a day.[38] Most of his time is spent on filtering and combining data, trying to verify claims and looking for additional information to rebut assertions. For his research, Van Linge taught himself Arabic via YouTube movies and learned to identify the various weaponry depicted in the online sources.[39] When a map is ready it is shared on Twitter.

Van Linge has spent several hours every day collecting data to update his map, using a smartphone, a laptop and an Internet connection. Once every two weeks he shared a new edition of "the situation in Syria" map on Twitter. According to Van Linge, it does not make sense to publish the map more frequently because the differences it would then show are too small.[40] Soon after Van Linge published his first map of Syria, his number of Twitter followers started to grow exponentially, from around two hundred before the first map to more than 52,000 in August 2019. The maps were retweeted and cited by news media like the *Huffington Post*, Lebanon's *Daily Star*, *The New York Times*, CNN and *Der Spiegel*, as well as by academic institutions such as the University of Texas.[41] Middle East experts have described Van Linge's maps as "among the most useful" and as "one of the best published on what's going on in Syria" and lauded the frequent updates.[42]

The moment Van Linge's maps were picked up by the international media coincided with the rise of jihadist group Islamic State. The feelings of disgust about their acts as well as fear about the territorial ambitions of the jihadist group created a need for clear information, which is something "the situation in Syria" maps provided. Many titles of the articles and reports on Van Linge's work refer to the jihadists and Islamic State and not to the broader issues of the Syrian civil war that Van Linge was trying to capture in his maps.

Today, Thomas van Linge is considered a Syria specialist as a result of his cartographic work. In 2018, together with Carla del Ponte, former member of the Independent International Commission of Inquiry on the Syrian Arab Republic of the UN Human Rights Council, Koos van Dam, scholar, diplomat and former Syria envoy, and others, Thomas van Linge was one of nine experts invited by the Dutch parliamentary committee on foreign affairs to speak about Syria.[43] The Dutch newspaper *Algemeen Dagblad* quotes Van Linge describing himself as a specialist during the meeting with the parliamentary committee.[44]

When asked about it, Van Linge refused to call his mapmaking a hobby.[45] "Hobby" would suggest it is a fun activity. He would rather call it a passion. On his Twitter account Van Linge describes himself as an activist: "Passionate about freedom, democracy, human rights and the preservation of our wildlife. Reporting on (and mapping out) wars, uprisings and conservation."[46] In the many video interviews with him from the summer of 2015, a flag of the Free Syrian Army features prominently on the wall of his bedroom in his parents' house where he worked on the maps. In the interviews he is clear about his ambitions to fight the Assad regime through media.[47] Notwithstanding this position, he indicates that he wants his maps to give a picture, as objective as possible, of what is going on in Syria.[48] The international media certainly seem to interpret them as such, presenting them unfiltered as neutral information. There is also criticism, but that is limited to indicating that more Thomas van Linges are needed, more maps from multiple perspectives, using diverse techniques to get a nuanced picture of the conflict.[49]

Van Linge's untitled first map of Syria from June 2013 shows a fragment of the Middle East with light gray for land, light blue for water and black lines for state borders. The map is tightly cropped around Syria, which has more detail than the surrounding area. The country has a yellow-brown sandy colored base with thin black lines for province borders, city names in a black sans serif typeface set in different sizes to indicate the size of the city. Placed next to the city names are circles in varying dimensions in accordance with the size of the city. The circles are colored to indicate the ruling party, green for the forces of the Syrian regime of President Bashar al-Assad, brown for the Free Syrian Army, an opposition army group, and blue for cities whose ruling party is contested. The map has no title, nor a map key.

The next Syria map published by Van Linge, seven months later in January 2014, is significantly different.[50] This new map has much more detail, including both area and city information in a more extensive and nuanced legend and, unlike the 2013 map, featuring a map key, a title ("the situation in Syria") and information about the author: Van Linge's name and Twitter handle @arabthomness (in 2018 changed to @ThomasVLinge). Between 18 January 2014 and 5 September 2018, Van Linge published sixty-five maps of Syria on Twitter, initially every fortnight

31 Van Linge, "IMPORTANT: map about the current situation in #syria."
32 Ray, "8 Deadliest Wars of the 21st Century."
33 "Thomas van Linge (20) brengt conflict in het Midden-Oosten in kaart."
34 Koens, "Deze 19-jarige Amsterdammer maakt kaarten van conflict Syrië."
35 Ricciardelli, "This Teenager Maps the Syrian War from His Bedroom."
36 Westcott, "The High School Student Who Maps ISIS's Lightning-Quick Advance."
37 "18-jarige kaartenmaker verovert de wereld."
38 Status on 11 August 2019: Thomas van Linge (@ThomasVLinge) has sent 42,500 Tweets since joining Twitter in January 2013. As a comparison, American president Donald Trump (@realDonaldTrump) sent 43,400 tweets, American media personality Kim Kardashian West (@KimKardashian) sent 29,800 tweets and American singer Katy Perry (@katyperry), who has the most followed Twitter account in the world, sent out 10,000 Tweets. All three joined Twitter in early 2009, four years before Van Linge started using the message platform. Twitter accounts @ThomasVLinge, @realDonaldTrump, @KimKardashian, @katyperry, accessed 11.08.2019.
39 "Thomas van Linge."
40 Egberts, "18-jarige Amsterdammer brengt feilloos Syrische strijdgroepen in kaart."
41 Kuntz, "The Dutch Teen Who Maps the Jihadists."
42 Westcott, "The High School Student Who Maps ISIS's Lightning-Quick Advance."
43 "Hoorzitting/rondetafelgesprek Nederlandse steun aan gewapende Syrische oppositie."
44 Keultjes, "Schone handen houden was onmogelijk in Syrië."
45 "Deze Nederlander van 20 volgt Islamitische Staat op de voet."
46 Twitter profile Thomas van Linge (@ThomasVLinge).
47 "A 19 ans, il est le meilleur cartographe du conflit syrien."
48 "Thomas van Linge."
49 De Weerd en Van Houtum, "Waarom je de macht van Islamitische Staat nooit in één kaart kunt vatten."
50 "The situation in Syria," map posted on Twitter (@ThomasVLinge), 18 January 2014.

Brush

Adobe Photoshop

Brush

Brush Presets

Brush Tip Shape

- Shape Dynamics
- Scattering
- Texture
- Dual Brush
- Color Dynamics
- Transfer
- Brush Pose
- Noise
- Wet Edges
- Build-up
- Smoothing
- Protect Texture

30	30	30	25	25	25	36
25	36	36	36	32	25	50
25	25	50	71	25	50	50
50	50	36	30	30	20	9
30	9	25	45	14	24	27
39	46	59	11	17	23	36
44	60	14	26	33	42	55
70	112	134	74	95	95	90
36	33	63	66	39	63	11
48	32	55	100	23	37	56
17	32	27	32	40	21	60
65	14	43	23	58	75	21
25	20	25	25	80	80	100
35	23	35	25	8	8	35
25	25	25	25	45	10	45
13						

Size: 30 px

Flip X   Flip Y

Angle: 0°

Roundness: 100%

Hardness: 0%

Spacing: 25%

New Brush Preset...

Clear Brush Controls
Reset All Locked Settings
Copy Texture to Other Tools

Close
Close Tab Group

Microsoft Paint

Brushes

(Right panel — Brush Presets)

Brush Presets

Size: 30 px

New Brush Preset...

Rename Brush...
Delete Brush

Text Only
Small Thumbnail
Large Thumbnail
Small List
Large List
Stroke Thumbnail

Preset Manager...

Reset Brushes...
Load Brushed...
Save Brushes...
Replaces Brushes...

Assorted Brushes
Basic Brushes
Calligraphic Brushes
DP Brushes
Drop Shadow Brushes
Dry Media Brushes
Faux Finish Brushes
M Brushes
Natural Brushes 2
Natural Brushes
Round Brushes with Size
Special Effect Brushes
Square Brushes
Thick Heavy Brushes
Wet Media Brushes

Close
Close Tab Group

	45
	14
	24
	27
	39
	46
	59
	11
	17
	23
	36
	44
	60
	14
	26
	33
	42
	55
	70
	112
25	134
50	74
	95
25	95
50	90
	36
	36
	33
	63
36	66
30	41
30	47
	10
9	27
	26
	41
9	38
	23
45	21
14	15
24	11
27	20
38	10
46	18
59	60
11	48
17	20
23	20
36	54
44	28
60	36
14	32
26	9
33	11

Comparison between the capabilities of the graphics software Microsoft Paint, used by Thomas van Linge, and Adobe Photoshop, the comparable tool used by specialists. Several functions of the tools, such as brushes, colors or lines, are compared by displaying the menus of the two softwares side by side. The extensive possibilities of Photoshop are striking, although it must be said that one seldom uses, or knows, all the possibilities.

Typesetting   Adobe Photoshop

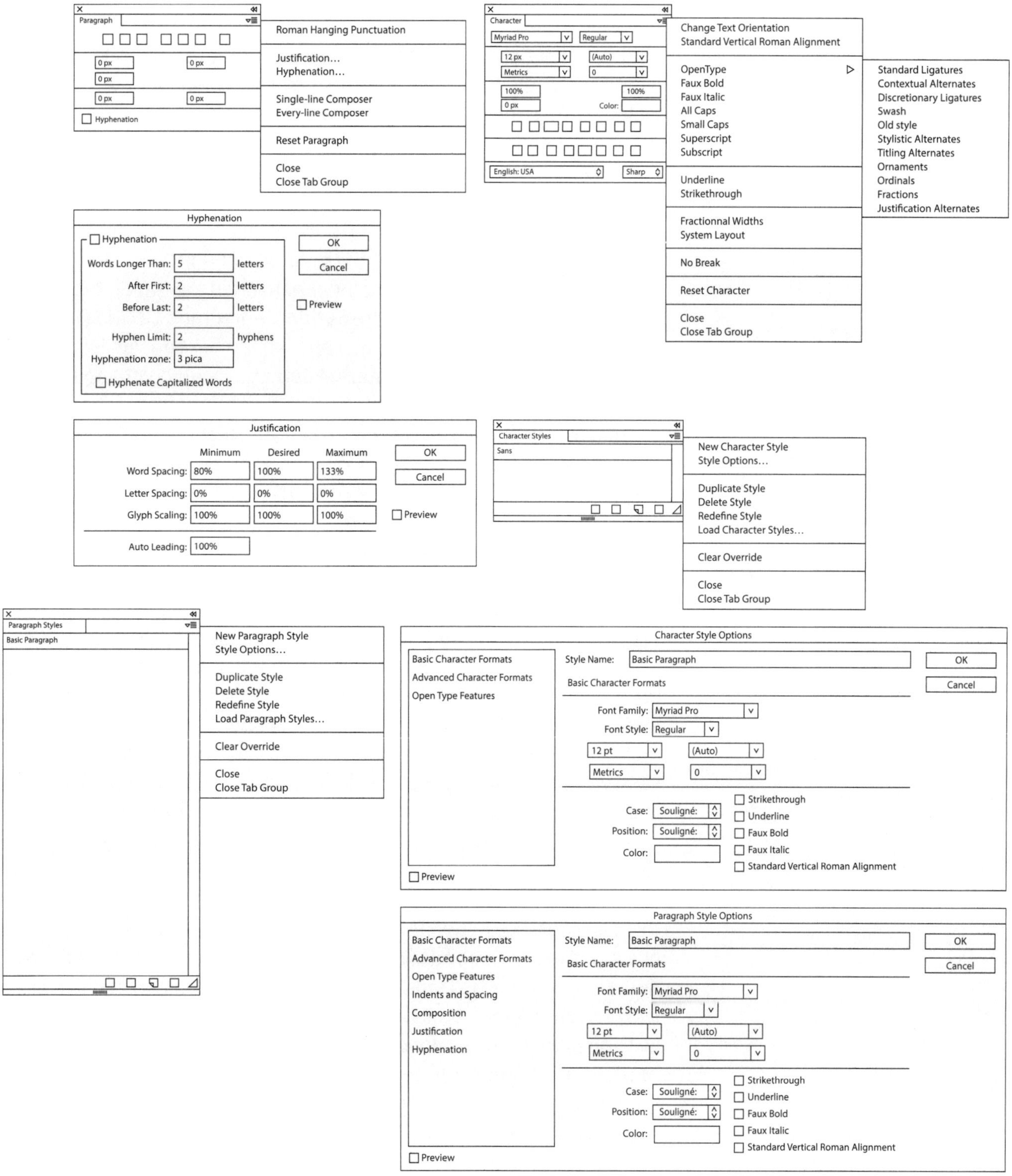

Paragraph panel menu:
- Roman Hanging Punctuation
- Justification…
- Hyphenation…
- Single-line Composer
- Every-line Composer
- Reset Paragraph
- Close
- Close Tab Group

Paragraph panel fields: 0 px, 0 px, 0 px, 0 px, 0 px, Hyphenation

Character panel:
Myriad Pro | Regular
12 px | (Auto)
Metrics | 0
100% | 100%
0 px | Color:
English: USA | Sharp

Character panel menu:
- Change Text Orientation
- Standard Vertical Roman Alignment
- OpenType ▷
- Faux Bold
- Faux Italic
- All Caps
- Small Caps
- Superscript
- Subscript
- Underline
- Strikethrough
- Fractionnal Widths
- System Layout
- No Break
- Reset Character
- Close
- Close Tab Group

OpenType submenu:
- Standard Ligatures
- Contextual Alternates
- Discretionary Ligatures
- Swash
- Old style
- Stylistic Alternates
- Titling Alternates
- Ornaments
- Ordinals
- Fractions
- Justification Alternates

Hyphenation dialog:
- Hyphenation
- Words Longer Than: 5 letters
- After First: 2 letters
- Before Last: 2 letters
- Hyphen Limit: 2 hyphens
- Hyphenation zone: 3 pica
- Hyphenate Capitalized Words
- OK
- Cancel
- Preview

Justification dialog:

	Minimum	Desired	Maximum
Word Spacing	80%	100%	133%
Letter Spacing	0%	0%	0%
Glyph Scaling	100%	100%	100%

Auto Leading: 100%
OK | Cancel | Preview

Character Styles panel: Sans

Character Styles panel menu:
- New Character Style
- Style Options…
- Duplicate Style
- Delete Style
- Redefine Style
- Load Character Styles…
- Clear Override
- Close
- Close Tab Group

Paragraph Styles panel: Basic Paragraph

Paragraph Styles panel menu:
- New Paragraph Style
- Style Options…
- Duplicate Style
- Delete Style
- Redefine Style
- Load Paragraph Styles…
- Clear Override
- Close
- Close Tab Group

Character Style Options dialog:
- Basic Character Formats
- Advanced Character Formats
- Open Type Features

Style Name: Basic Paragraph
Basic Character Formats
Font Family: Myriad Pro
Font Style: Regular
12 pt | (Auto)
Metrics | 0
Case: Souligné:
Position: Souligné:
Color:
- Strikethrough
- Underline
- Faux Bold
- Faux Italic
- Standard Vertical Roman Alignment
- Preview
OK | Cancel

Paragraph Style Options dialog:
- Basic Character Formats
- Advanced Character Formats
- Open Type Features
- Indents and Spacing
- Composition
- Justification
- Hyphenation

Style Name: Basic Paragraph
Basic Character Formats
Font Family: Myriad Pro
Font Style: Regular
12 pt | (Auto)
Metrics | 0
Case: Souligné:
Position: Souligné:
Color:
- Strikethrough
- Underline
- Faux Bold
- Faux Italic
- Standard Vertical Roman Alignment
- Preview
OK | Cancel

Microsoft Paint

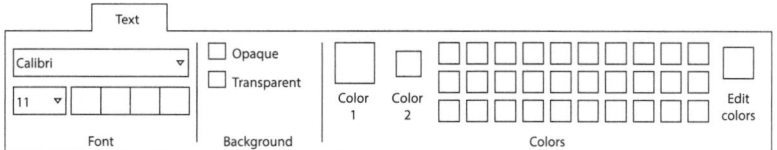

Calibri
11
Font

Opaque
Transparent
Background

Color 1 | Color 2
Edit colors
Colors

but later less frequently, all following the same design, with the exception of a few small improvements. For example, the maps published from June 2014 onwards use a legend that includes a second lighter shade to identify areas with a lower population density.[51] The lighter color is used for deserts and less densely populated areas, the darker and brighter colors are used for the more densely populated northern edge and western part of the country, where Syria's largest cities Aleppo, Damascus and Homs are situated as well as the zone around the River Euphrates that crosses the country diagonally. Another change is the terminology used in the map's legend. The category "Kurds," in reference to the stateless ethnic group, becomes "Rojava"[52] on the map of 1 January 2015, the name for the autonomous region in northeastern Syria, to be changed on 20 February 2017 into "Federation Forces,"[53] following the renaming of the region to Democratic Federation of Northern Syria by the Syrian Democratic Council in Rmelan in late December 2016. Comparing one map with the next on Van Linge's Twitter feed is reading the mapmaker's "internal debate," in which changes in the political situation or other insights that informed the cartographic design can be observed.

I already mentioned that the embedding of a map in a public debate accentuates its post-representational aspects. When it is shared, distributed and becomes part of a conversation, a map becomes processual. In the case of "the situation in Syria" map, four processual debates can be identified. First, there is the Twitter feed, which is Van Linge's main source of communication. The many Tweets, Retweets and Twitter Threads in Van Linge's feed provide context to the map. Simultaneously, it is a source of input, a platform for discussion and a means to give accountability. The second debate is the internal one as described above, in which the earlier editions of the map remain available for comparison to reveal the contemplations of the producer. The third debate is Thomas van Linge's public appearances outside his Twitter feed, such as giving interviews about his maps and talking about his methods and motivation. The fourth debate is the visible traces on the map of earlier versions. Of the four, this debate is the hardest to discern and also one that some might not opt to include, but to me it is an important one, as I will explain below.

Whereas nowadays most maps are made using vector graphics editors, "the situation in Syria" map is produced with a pixel graphics editor. I will elaborate later on the exact tools Van Linge uses and why that is relevant, for now it is important that the difference between a pixel- and vector-based graphic is that the former is built up of many small squares that when zoomed in look jagged, pixelated, and the latter are defined in terms of precise points connected by mathematically defined lines and curves that remain smooth even when zoomed in very closely. A popular pixel-graphics editor is Adobe Photoshop that is used to edit, to "photoshop," raster-based imagery like photographs. Unlike pixel-based images, vector-based illustrations are infinitely scalable. This format is often used for typefaces, page layouts and architectural drawings. The main difference between the two types of tools in cartographic production is that maps made using a vector graphics editor can be more precise and their use more diverse in terms of the size in which they are presented, as their information is scalable without a loss of graphic quality.

One of the tools used to edit "the situation in Syria" map, I presume, is a so-called Paint Bucket tool. This tool fills a selected area with a color. Depending on the settings, it can either permeate an entire field or ignore certain elements, like a text. This ignoring is achieved by specifying a certain tolerance of color values in the settings. If the tolerance is set to 0, only one specific color value will be filled in the selected area. The higher the tolerance, the more pixels will be covered. A text in a pixel-based image does not only consist of pixels in a single color, but also of several shades of that color to give the impression of smooth curves, rather than a hard jagged outline. If a black text is placed on a red background, the pixel editor will create the appearance of soft curves by making the pixels on the edge of the letters in a color value between black and red. If the background color is then adjusted to gray without using the correct paint bucket settings, the black text retains some dark-red pixels around the letters. These traces are clearly discernible when viewed on a pixel level, but in a normal viewing setting they are only slightly visible. Still, on Van Linge's Syria map, in areas that saw many changes during the civil war, the traces are a visual clue of eventful times.

Take, for instance, the city of Palmyra. In a two-year period it was first ruled by the loyalists of the Assad regime, then controlled by Islamic State, only to return to loyalist rule. On the map of 20 February 2017, the word Palmyra has a red aura, set against a light-gray background indicating the control of the Islamic State. This trace of the red background color still indicates the former control of the area by the Assad regime.[54] The red aura is also visible on the map of 1 June 2014, this time against a pink background.[55] Previously, the background around Palmyra had been colored bright red, but from this map onwards Van Linge makes a distinction between low- and high-density areas: sparsely populated areas are shown in a lighter shade; densely populated areas have a darker shade. The word Palmyra, while referring to a populous city, is placed on the map below a big dot in the empty desert area. The red aura that in 2017 is a trace of a previous regime, is a trace of a different cartographic approach in 2014.

To me, these traces constitute a fourth processual debate in "the situation in Syria" map. The digital residue of previous editions of the map is a visual cue of an ongoing negotiation between the map, the circumstances that it depicts, the cartographic language in which this is done and the technology that is used to produce it. It also provides an insight in the learning process of the mapmaker. Van Linge introduces new elements on the map as he gains a better understanding of the situation on the ground and, perhaps, advances his cartographic knowledge. The great strength of the debate of traces is that, unlike the other public debates, it reveals itself in a single map. But only for those who are willing and capable to dissect it on such a fundamental level. The Twitter debate and the internal debate disclose themselves as a sequence, a follow-up of messages or a comparison of consecutive editions of the map that do not show up if one studies an individual image.

One could argue that traces of previous versions in "the situation in Syria" map are digital imperfections resulting from a lack of skills of the mapmaker or the use of an inadequate tool. They would probably not be present if Van Linge had more knowledge and experience with graphic production. Van Linge's statement in an interview with American news magazine and website Newsweek that he is "not very sophisticated with computers" confirms this.[56] I see the traces as an unintentional quality that provide the map with an additional layer of information. More generally, to me such imperfections grant an air of sincerity to the graphic

51   "The situation in Syria," map posted on Twitter (@ThomasVLinge), 1 June 2014.
52   Ibid., 1 January 2015.
53   Ibid., 20 February 2017.
54   Ibid.
55   Ibid., 1 June 2014.
56   Westcott, "The High School Student Who Maps ISIS's Lightning-Quick Advance."

Color　　　　　　　　Adobe Photoshop

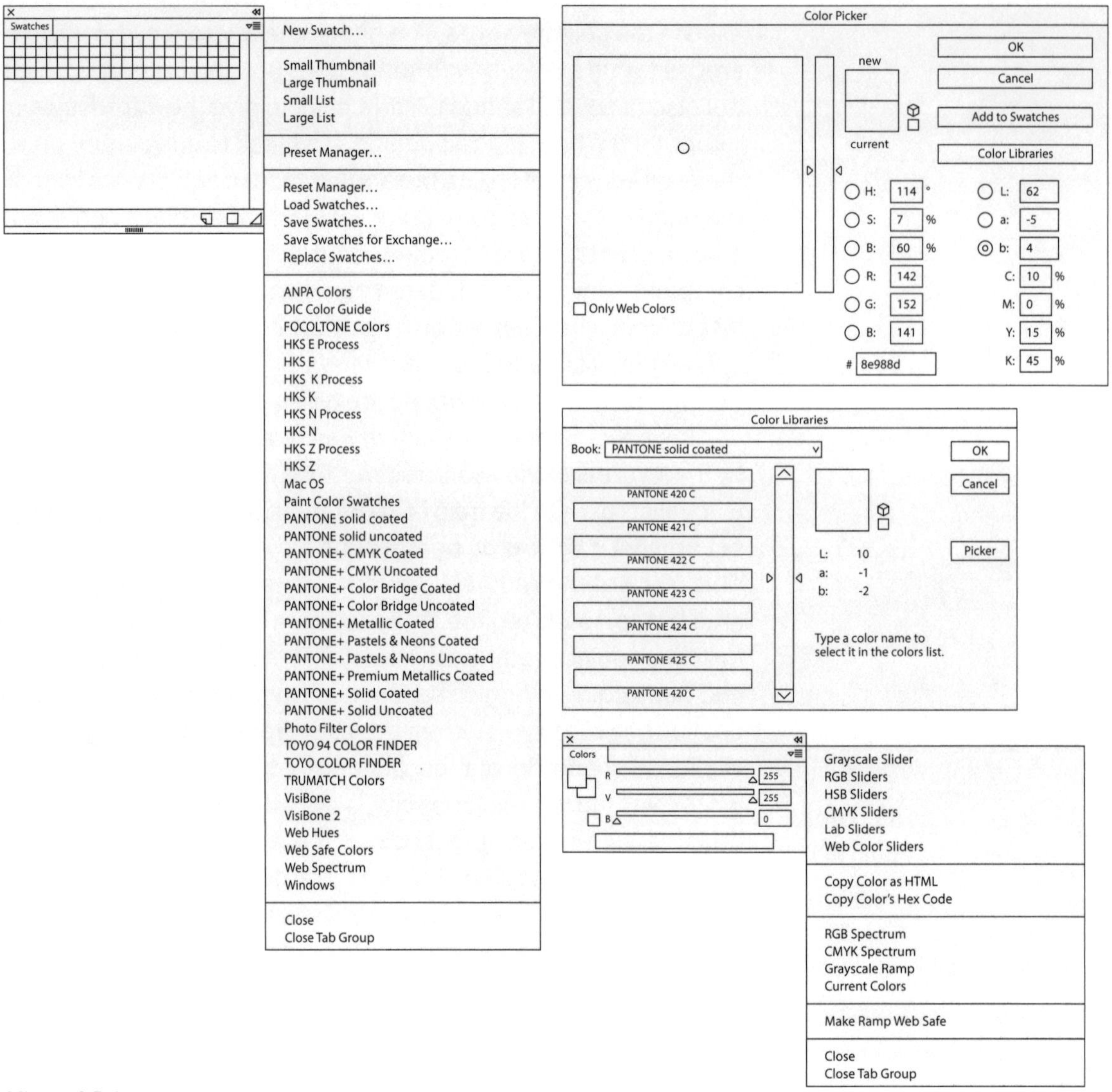

New Swatch…	
Small Thumbnail	
Large Thumbnail	
Small List	
Large List	
Preset Manager…	
Reset Manager…	
Load Swatches…	
Save Swatches…	
Save Swatches for Exchange…	
Replace Swatches…	

ANPA Colors
DIC Color Guide
FOCOLTONE Colors
HKS E Process
HKS E
HKS K Process
HKS K
HKS N Process
HKS N
HKS Z Process
HKS Z
Mac OS
Paint Color Swatches
PANTONE solid coated
PANTONE solid uncoated
PANTONE+ CMYK Coated
PANTONE+ CMYK Uncoated
PANTONE+ Color Bridge Coated
PANTONE+ Color Bridge Uncoated
PANTONE+ Metallic Coated
PANTONE+ Pastels & Neons Coated
PANTONE+ Pastels & Neons Uncoated
PANTONE+ Premium Metallics Coated
PANTONE+ Solid Coated
PANTONE+ Solid Uncoated
Photo Filter Colors
TOYO 94 COLOR FINDER
TOYO COLOR FINDER
TRUMATCH Colors
VisiBone
VisiBone 2
Web Hues
Web Safe Colors
Web Spectrum
Windows

Close
Close Tab Group

Color Picker

OK
Cancel
new
Add to Swatches
current
Color Libraries

☐ Only Web Colors

○ H:	114	°	○ L:	62
○ S:	7	%	○ a:	-5
○ B:	60	%	◉ b:	4
○ R:	142		C:	10 %
○ G:	152		M:	0 %
○ B:	141		Y:	15 %
#	8e988d		K:	45 %

Color Libraries

Book: PANTONE solid coated ∨

PANTONE 420 C
PANTONE 421 C
PANTONE 422 C
PANTONE 423 C
PANTONE 424 C
PANTONE 425 C
PANTONE 420 C

OK
Cancel
Picker

L:　10
a:　-1
b:　-2

Type a color name to
select it in the colors list.

Colors

R ———— 255
V ———— 255
☐ B △—— 0

Grayscale Slider
RGB Sliders
HSB Sliders
CMYK Sliders
Lab Sliders
Web Color Sliders

Copy Color as HTML
Copy Color's Hex Code

RGB Spectrum
CMYK Spectrum
Grayscale Ramp
Current Colors

Make Ramp Web Safe

Close
Close Tab Group

Microsoft Paint

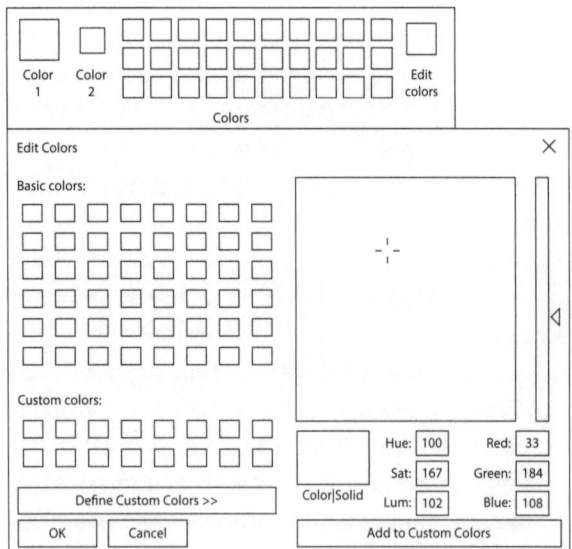

Color 1　Color 2　　　　　Edit colors

Colors

Edit Colors　　　　　　　✕

Basic colors:

Custom colors:

Define Custom Colors >>

OK　　Cancel

Color|Solid

Hue:	100	Red:	33
Sat:	167	Green:	184
Lum:	102	Blue:	108

Add to Custom Colors

Lines   Adobe Photoshop   Microsoft Paint

Layers   Adobe Photoshop

Microsoft Paint has no layer option

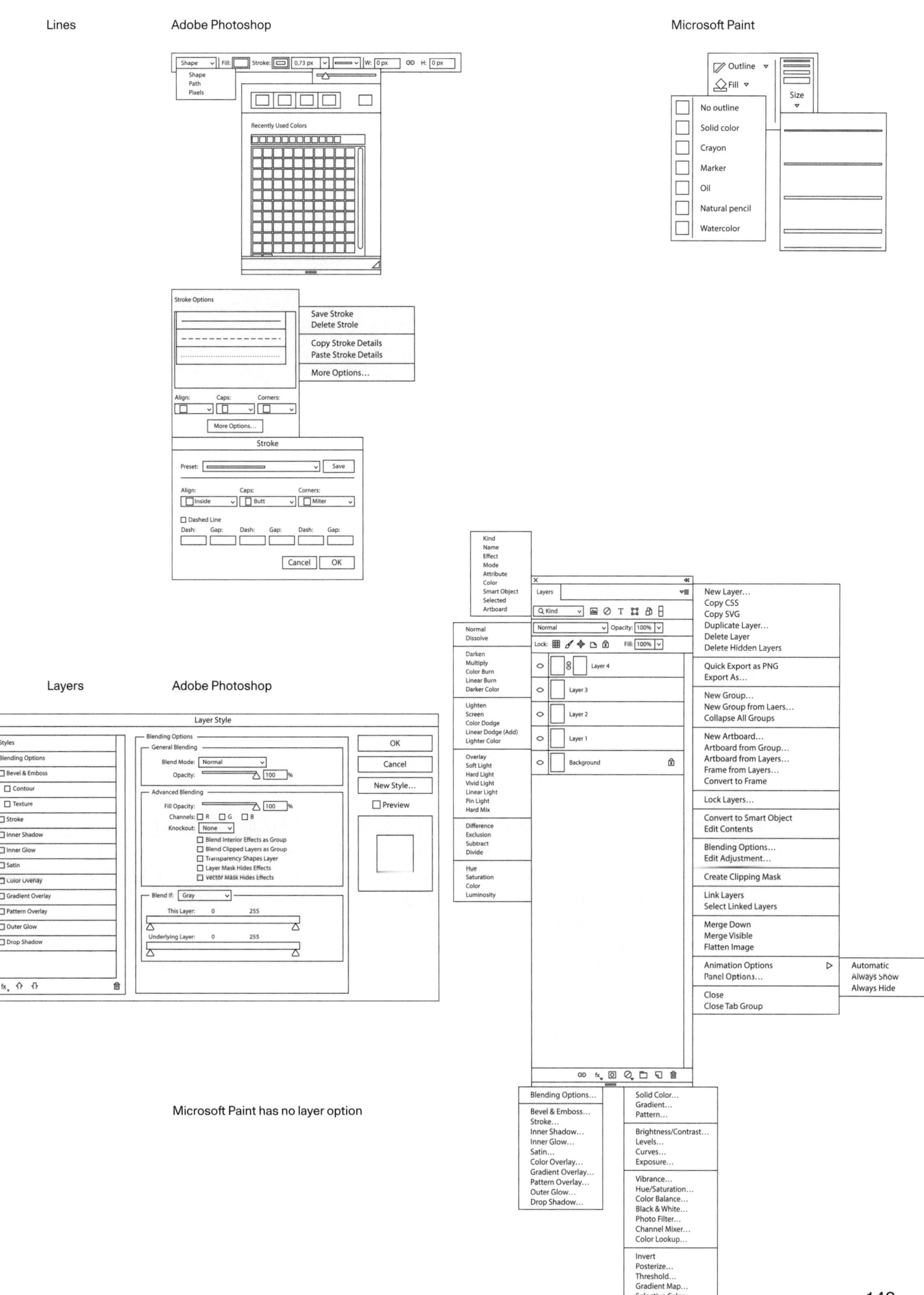

product. As if with the lack of skills, also the power to rhetorically manipulate and influence the user is absent. Besides, any shortcomings of design competence do not prevent the mapmaker from adequately expressing her intentions. I agree with Offenhuber when he concludes that the lack of refinement and raw appearance of the work of amateur conflict mapmakers is "not unintentional; it follows a visual logic that serves the purpose of presenting evidence."[57]

With the map tightly cropped around the country, the title positioned in the top center and the legend at the bottom, both set in a serif typeface, light blue for water and bright colors for the different parties that control the country, "the situation in Syria" map has the appearance of traditional cartography. Van Linge's map differs from a printed atlas in that it is embedded in several public debates, it is shared, distributed and part of an ongoing dialogue and open to being questioned. Another difference is that Van Linge's maps contain traces of previous versions that give the discourse a visual presence on the map, albeit hardly perceptible. Both aspects highlight and emphasize the post-representational processual nature of the map. In addition to these two, there is a third aspect in which "the situation in Syria" map distinguishes itself from representational cartography and that is how it is produced.

Van Linge's maps are not made with a specialist tool, but by using software called Microsoft Paint, a pixel graphics editor with limited possibilities that is part of the Microsoft Windows operating system.[58] To better explain why this is an important aspect to consider, I will explain how graphics are produced in a professional context, how graphic designers are trained and why Van Linge's choice to use Microsoft Paint is undermining both of these.

Insurgency Tactics

In the past few years I conducted a series of workshops in Switzerland and the Netherlands entitled "Atlas of Design Tools."[59] The aim of the workshops was to investigate political, social, environmental and economic aspects of the design tools the participants were using in their daily practice. The critical inquiry resulted in research questions such as who created the tools, when and where they were invented and produced, what materials and means were used to produce them, under what conditions they were made, the costs of their production, the ownership of the technology behind them, and who it is that makes the profits. The outcome of the research, translated into visual formats like maps, timelines and diagrams, was collected into a single book at the end of the workshop: the atlas of design tools.

At the start of the workshops the participants were asked to list their most-used design tools. Out of more than one hundred participants, a large majority mentioned Adobe software as the technology they favored. The top ten at each workshop consistently contained, and often in the upper positions, Adobe Photoshop, Adobe Illustrator and Adobe InDesign, respectively the pixel graphics editor, the vector graphics editor and desktop publishing and typesetting application of American software company Adobe. The applications listed above are part of the Adobe Creative Cloud, a set of software tools aimed at graphic design, video editing, web development and photography that is the industry standard in many creative fields.

Two notions coming from the workshops' critical investigations into Adobe software are worth mentioning. First, the dominance of Adobe in the field of creative tools. There are other options for Adobe software, both commercial and open-source, but these are few and their use is limited. Furthermore, the alternatives concern single software tools in the Creative Cloud package. But I would claim that we should not look at the individual tools but at the package as a whole. The set of Adobe softwares offers the possibility to seamlessly transport designed elements from one application to the next. To Russian/American author and scholar Lev Manovich, this crossing over from one tool and medium to the next is a typical practice of what he calls the "software era."[60] Working on a project, a designer will use several softwares and take a design element and work on it in a variety of tools. Manovich states that "'import,' 'export' and related functions and commands ... are more important than the individual operations [the] programs offer."[61] In other words, when dealing with graphics and video editing tools we should not consider the individual tool, but the group of tools as a whole. On that level there is no alternative for the Creative Cloud, there is no set of softwares that is so complete and offers such an easy exchange between tools as Adobe does. Considering all this, the earlier mentioned term *industry standard* feels like an understatement to describe the position of Adobe's software; *monopoly* would be a more appropriate term.

A second notion that came from the workshops is a critique of Adobe's economic model. Since 2014, Adobe's software tools are available in subscription format only. The monthly costs amount to around €36 for an individual tool like Adobe Photoshop, or €90 for the full set of tools in Adobe Creative Cloud, meaning that if one were only using the software to make one map per two weeks, like Thomas van Linge does, the costs would amount to €18 per map if only Photoshop is used or €45 per map if a combination of tools is utilized.[62] Adobe's economic model forces the producers of graphics to work in a specific way. The cost of the tool is only justified if it is used in a specialized practice.

In my own graphic design practice I also notice the dominance of Adobe. Certain publishers or printers only work with Adobe software. When working with these parties, one is obliged as a designer to use these specific tools. I also experience the dominance of the American software company in another way.

A few years ago I started to research the current state of the field of graphic design. In theory and writings about the discipline, I could not find descriptions of the aspects I was experiencing in my practice, like the shifting position of the designer. So I started researching it myself. This book is one example, the workshops I described above another. In yet another part of the research, an investigation done in my studio, we looked at the applications we receive from students who are looking for an internship. We looked at the CVs and portfolios of around two hundred applicants in the years 2016 and 2017 to study the education and first professional activities of a new generation of graphic designers. The applications came from all over the world, the majority from Europe, with the highest number of applicants coming from France. Of the applicants who listed skills on their CV, all were proficient in the Adobe Creative Cloud tools. Only a handful

57 Dietmar Offenhuber, "Maps of Daesh: The Cartographic Warfare Surrounding Insurgent Statehood."

58 In almost all articles and news reports about Van Linge's maps, this fact is featured prominently. A contrast is created between on the one hand, a teenager living with his parents, 3,000 km from Syria, making maps using Microsoft Paint as opposed to war violence, the precision of the maps and their usage by prominent international news media.

59 The workshops took place at departments of graphic design, both at bachelor and master level, of École Cantonale d'art de Lausanne (2016), Zurich University of the Arts (2017), Design Academy Eindhoven (2018) and Bern University of the Arts (2018). The supervision of all but one of the workshops was done in collaboration with Dimitri Jeannottat.

60 Manovich, *Software Takes Command*.

61 Ibid., 306.

62 "Ontdek de Creative Cloud-ervaring."

1 California Institute of the Arts, Valencia
2 Rhode Island School of Design, Providence
3 Universidade Europeia, Lisboa
4 Escola Superior de Artes e Design, Porto
5 Glasgow School of Art, Glasgow
6 Leeds Beckett University, Leeds
7 Central Saint Martins, London
8 Kingston University, London

9 Escuela Universitaria de Diseño et Ingeniería de Barcelona, Barcelona
10 École de Communication Visuelle, Nantes
11 Atelier de Sèvres, Paris
12 École Estienne, Paris
13 École de Communication Visuelle, Paris
14 Institut Supérieur de Communication et de Publicité, Paris
15 École nationale supérieure des Arts Décoratifs, Paris

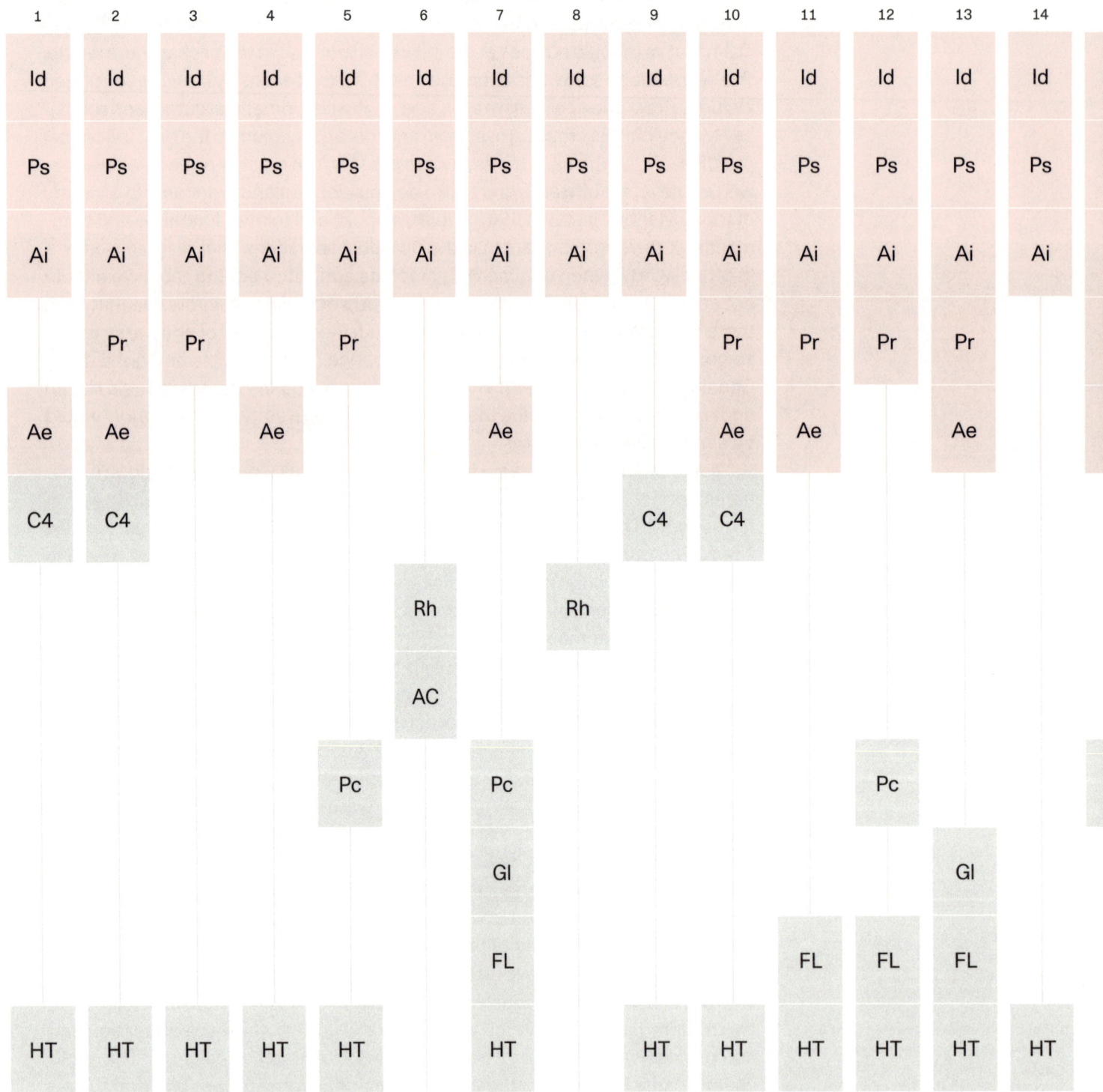

Id Adobe InDesign is a layout and page design software for print and digital media developed by Adobe Inc. (US) and released in 1999.
Ps Adobe Photoshop is an imaging and graphic design software developed by Adobe Inc. (US) and released in 1990.
Ai Adobe Illustrator is a vector graphics software developed by Adobe Inc. (US) and released in 1987.
Pr Adobe Premiere is a video editing software for film, TV and the web developed by Adobe Inc. (US) and released in 1991.
Ae Adobe After Effects is a motion graphics and visual effects software developed by Adobe Inc. (US) and released in 1994.

C4 Cinema 4D is a 3-D modeling, animation, motion graphic and rendering software developed by MAXON (Germany) and released in 1990.
Rh Rhinoceros is a 3-D computer-aided design modeling software developed by Robert McNeel & Associates (US) and released in 1994/1998.
AC AutoCad is a computer-aided design and drafting software developed by Autodesk Inc. (US) and released in 1982.
Pc Processing is an open-source graphical library and integrated development environement developed by Casey Reas and Ben Fry (US) in 2001.
Gl Glyphs is a font editor developed by Georg Seifert (Germany) and released in 2011.
FL FontLab is a font editor developed by SoftUnion Ltd (Russia)/Pyrus North America Ltd (US) and released in 1993.
HT HTML is the standard markup language for documents designed to be displayed in a web browser, developed by Tim Berners-Lee (UK) in 1992.

▢ Adobe
▢ Other developer

16	17	18	19	20	21	22	23	24	25	26	27	28	29	30
Id	Id	Id	Id	Id	Id	Id	Id	Id	Id	Id	Id	Id	Id	Id
Ps	Ps	Ps	Ps	Ps	Ps	Ps	Ps	Ps	Ps	Ps	Ps	Ps	Ps	Ps
Ai	Ai	Ai	Ai	Ai	Ai	Ai	Ai	Ai	Ai	Ai	Ai	Ai	Ai	Ai
Pr	Pr	Pr	Pr	Pr		Pr		Pr			Pr		Pr	Pr
	Ae	Ae	Ae	Ae		Ae		Ae	Ae		Ae	Ae	Ae	Ae
		C4		C4		C4	C4			C4				
									Rh	Rh			Rh	
					AC				AC	AC				
		Pc		Pc							Pc	Pc		
			Gl	Gl		Gl							Gl	
	FL	FL	FL										FL	
	HT	HT	HT	HT	HT	HT	HT	HT	HT	HT	HT	HT		HT

Overview of the software skills listed in the curricula vitae of students who applied for an internship at graphic design studio SJG, sorted by school. These data are the result of a study of about 200 internship applications in the years 2016 and 2017. The applications came from all over the world, the majority from Europe, with the highest number of applicants coming from France. Of the applicants who listed skills on their CV, all were proficient in the Adobe Creative Cloud tools.

mastered other graphics editors. These new designers almost exclusively mastered Adobe software. So even if I am not working with publishers and printers, if I was working on self-initiated projects that I would self-produce then I would still be dependent on Adobe tools if I wanted to collaborate with other designers.

The predominant position of one software package in a field is not limited to Adobe Creative Cloud in the field of graphic production. Think for instance of the role of Microsoft Office in the academic field: Microsoft Word to write papers or research proposals, Microsoft PowerPoint to give presentations at conferences or in education, and Microsoft Excel to do quantitative research, make a diagram, or keep a budget. The point I want to make is that the dominance of a tool in a discipline leads to a certain specialized practice. Some have even claimed that software tools also have an impact on the output. American statistician, scholar and writer on information design Edward R. Tufte has argued that Microsoft Powerpoint forces users to create presentations that use "an intensely hierarchical single-path structure as the model for organizing every type of content" that "turns information into a sales pitch and presenters into marketers."[63]

Bearing the above in mind, it is clear that not using the dominant tool, the seemingly only tool in a field, is an act of rebellion. That is at least how I see Thomas van Linge's use of Microsoft Paint to make a map. Van Linge found a way to escape the default. Not by the appropriation of a tool, as many designers strive for, but by denying it and instead using a "nontool." In doing so, he developed a graphics editing practice that escapes the model that the "industry standard" seems to demand.

Much more can be said on the subject of software. It is a topic that is highly relevant and explored in depth in the field of software studies by scholars like the aforementioned Lev Manovich and Matthew Fuller, writer, artist and professor of cultural studies at Goldsmiths University of London. In the context of this research I have chosen to focus on the impact of software on practice. Tools rooted in a certain production model demand a certain type of specialized practice and a certain type of specialized education of its practitioners. They also seem to foster certain kinds of outcomes.

In Chapter 2, I used the military term *friendly fire* to describe the impact of the democratization of design tools on the position of the designer, an attack by a force on one's own or neutral units while attempting to attack the enemy. The digitization of the production process empowered the designer, but also made it possible for outsiders to enter the field and subsequently render the designer obsolete. In the digital age nonspecialists have access to specialist tools. In the case of Thomas van Linge's practice, there is a slight difference as he uses nonspecialist tools to produce his work. To describe Van Linge's practice I would like to use another military concept: insurgency tactics. In military terms these are the actions of rebels against an established government, often involving improvised and homemade weaponry. To stay in the military analogy, if Adobe Creative Cloud is a state-of-the-art fighter jet, then Microsoft Paint is a Molotov cocktail or baseball bat: inexpensive, employable without training, not very subtle, but highly effective.

Postscript

Three years after the publication of Tomás López's *Atlas geográfico de España*, the Peninsular War started. In this military conflict between Napoleon Bonaparte's French empire and Bourbon Spain assisted by the United Kingdom and Portugal, the French occupied Spain.[64] The most recent geographic source available was López's atlas. But soon both Napoleon's and Wellington's armies found out that it lacked precision and started making their own maps of Spain.[65] The errors were due to the nontopographical surveying method López had used. He had learned the desk cartography method from one of the eighteenth century's most prestigious cartographers, Frenchman Jean Baptiste Bourguignon d'Anville (1697–1782), but López applied it with less rigor and did not give precise instructions to the village priests who carried out the surveys. The use of the maps in a situation of conflict had exposed the flaws of López's expertise.

The quality of Thomas van Linge's maps established his reputation as a specialist. Not only on Syria, but more generally on reporting and mapping current wars and uprisings, in line with the various topics covered in his Twitter feed. While writing this text in August 2019, Van Linge, who is currently a political science student, is in Hong Kong reporting on the antiextradition bill protests for the news program of Dutch broadcaster RTL.[66] Shifting from the mapmaking of situations to mapping conflicts, Thomas van Linge has become a specialist.

Although the above paragraphs might seem to suggest otherwise, the maps of the two Thomases are not that different. Both use people on the ground to gather data and map a site. Both compare a variety of sources to extract information and make a map. The specialist label that might distinguish one of the two seems arbitrary. What makes the difference is whether the map gains authority in its use.

Conclusion

This chapter looked into the practices of amateur conflict mapmakers, nonspecialists who lack training, knowledge and skills but who instead have knowledge of, or access to, a site or are willing to make inexhaustible efforts to collect and compare data and convert it into a map. The work produced by these practices is more raw and unsophisticated than those of specialist practices, which gives their work an air of honesty; the absence of sophistication equals the absence of rhetorical manipulation.

What interests me about the work of amateur conflict mapmakers is that in a way they produce an antirepresentational cartography. The premise of representational cartography is that the world can be objectively known and truthfully mapped. One of the aims of cartography, according to this approach, is to improve the effectiveness of a map. The design of a map, the relationship between a map's content and its graphic containers, should be carefully balanced. Amateur conflict mapmakers lack the skills, knowledge and perhaps even the interest to make the kind of sophisticated map that representational cartography strives for. Given the nature of their subject matter and the—mostly online—debates they are part

63   Tufte, *The Cognitive Style of PowerPoint*, 4.
64   Francisco Goya's famous painting *The Third of May 1808* (1814) depicts a key event of the Peninsular War: the uprising of citizens of Madrid against the French army.
65   San-Antonio-Gómez, Velilla and Manzano-Agugliaro, "Tomás López's Geographic Atlas of Spain in the Peninsulan War: A Methodology for Determining Errors."
66   "Botsingen tussen politie en betogers op bezet vliegveld Hongkong."

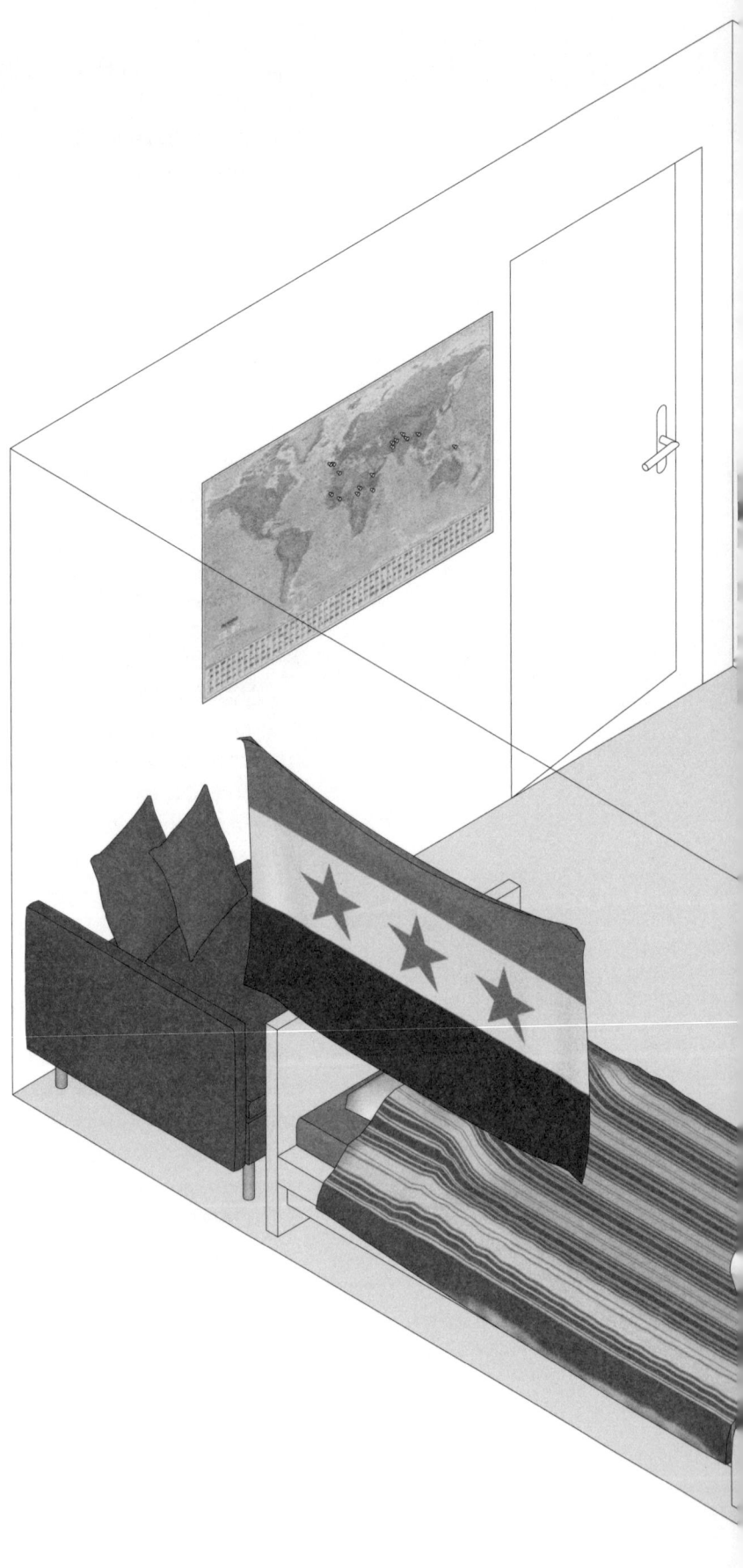

Thomas van Linge, Amsterdam

Sources: "18-jarige kaartenmaker verovert de wereld"; "A 19 ans, il est le meilleur cartographe du conflit syrien"; "Deze Nederlander van 20 volgt Islamitische Staat op de voet"; Koens, "Deze 19-jarige Amsterdammer maakt kaarten van conflict Syrië"; Ricciardelli, "This Teenager Maps the Syrian War from His Bedroom."

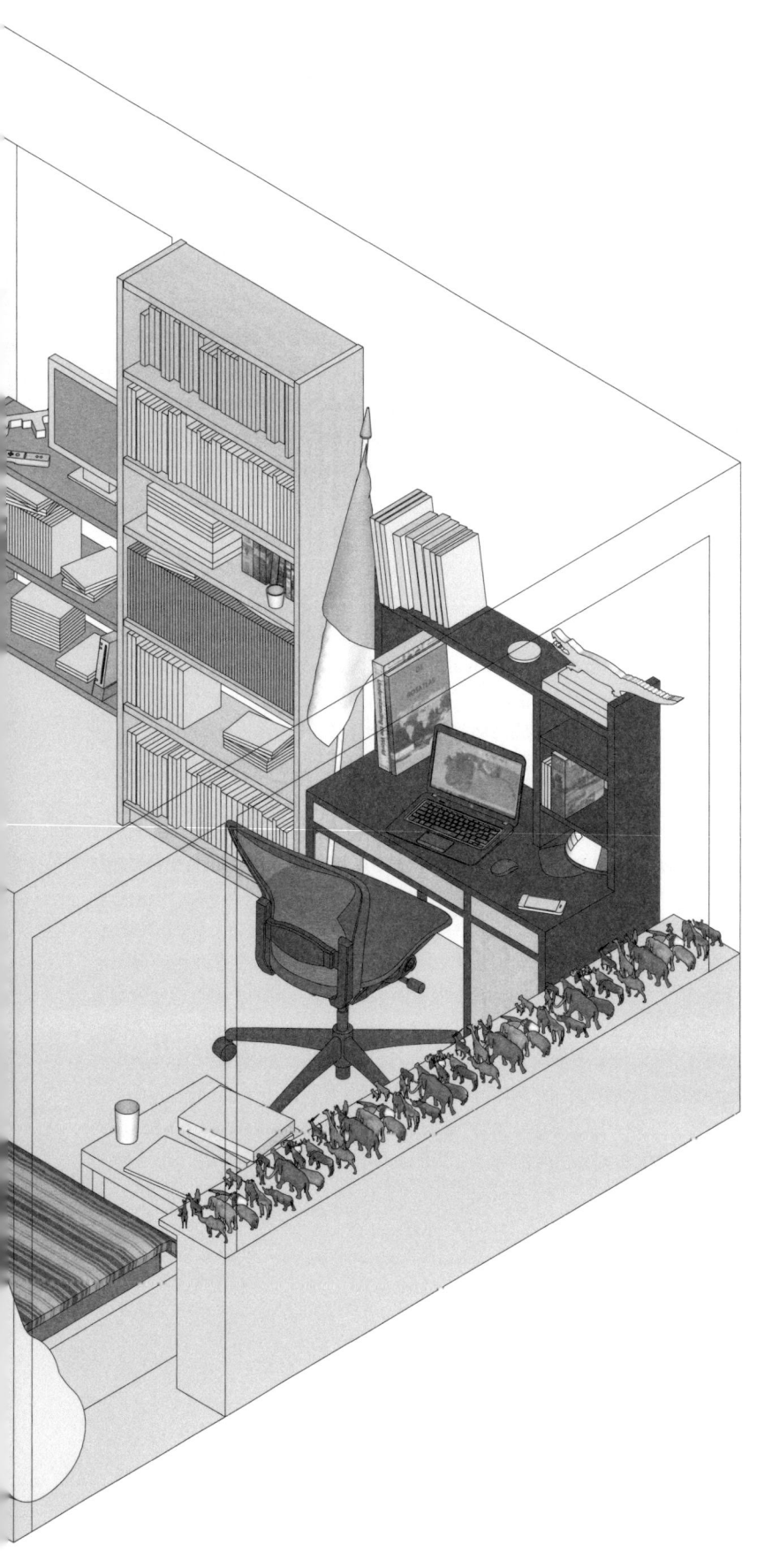

of, amateur conflict mapmakers put more emphasis on the process of unearthing data, on showing what otherwise would be hidden, than on the representation of information as an image, on mapmaking. Their maps are visibilizations, rather than visualizations. In other words, the work of amateur conflict mapmakers highlights the process of mapping rather than the act of mapmaking. The terms *visualization* and *visibilization* highlight the intentions of the mapmakers, rather than the actual use of the maps. Amateur conflict mapmaker Thomas van Linge's maps, for instance, are adopted as truthful representations by major news media.

The maps of amateur conflict mapmakers might lack certain visual qualities. Their work, however, has other merits, especially when approached from a post-representational perspective. In post-representational cartography a map is regarded as a process rather than as a product. The full process of creating, editing, producing, distributing and using is never finished. A map is reproduced again and again, every time a user engages with it. When considering the map as a process, it becomes important to look beyond the map as a graphic product, to how it is shared, distributed, whether it is part of an ongoing dialogue and is open to being questioned. In other words, not so much the visual strategies, but whether it is part of a public debate determines the accountability of the map. A map can be embedded in several debates. These can vary from open dialogues on social media where maps are shown amid their sources along with previous versions and discussions about claims made in the map, to more internalized debates that show the process and considerations of a mapmaker. Most of these public debates become apparent when comparing various maps, or looking at the map in the context of a conversation. There is also a discourse that shows up on the map itself. Digital residues of previous versions of the map leave a subtle trace on a map to constitute a public debate.

Another processual aspect that is not visible in their maps, but has a significant impact on the practice of amateur mapmakers, is how the work is produced. Rather than following developments in practices, tools seem to shape the practices in which they are used. Specialized software generates a specialized practice because of the economic model employed by software manufacturers, the demands of an industry that uses the software as an exchange format, and the curricula of educational institutes. By means of improvisation and the use of generic tools, amateurs may create different kinds of graphics editing practices not based on specialization. Specialized design practices can learn from nonspecialists to adopt their strategies to reformulate one's practice, improvise, use nonspecialist tools and engage one's work and oneself in public debates to "walk and measure" the world.

Thomas van Linge, Amsterdam

Sources: "18-jarige kaartenmaker verovert de wereld"; "A 19 ans, il est le meilleur cartographe du conflit syrien"; "Deze Nederlander van 20 volgt Islamitische Staat op de voet"; Koens, "Deze 19-jarige Amsterdammer maakt kaarten van conflict Syrië"; Ricciardelli, "This Teenager Maps the Syrian War from His Bedroom."

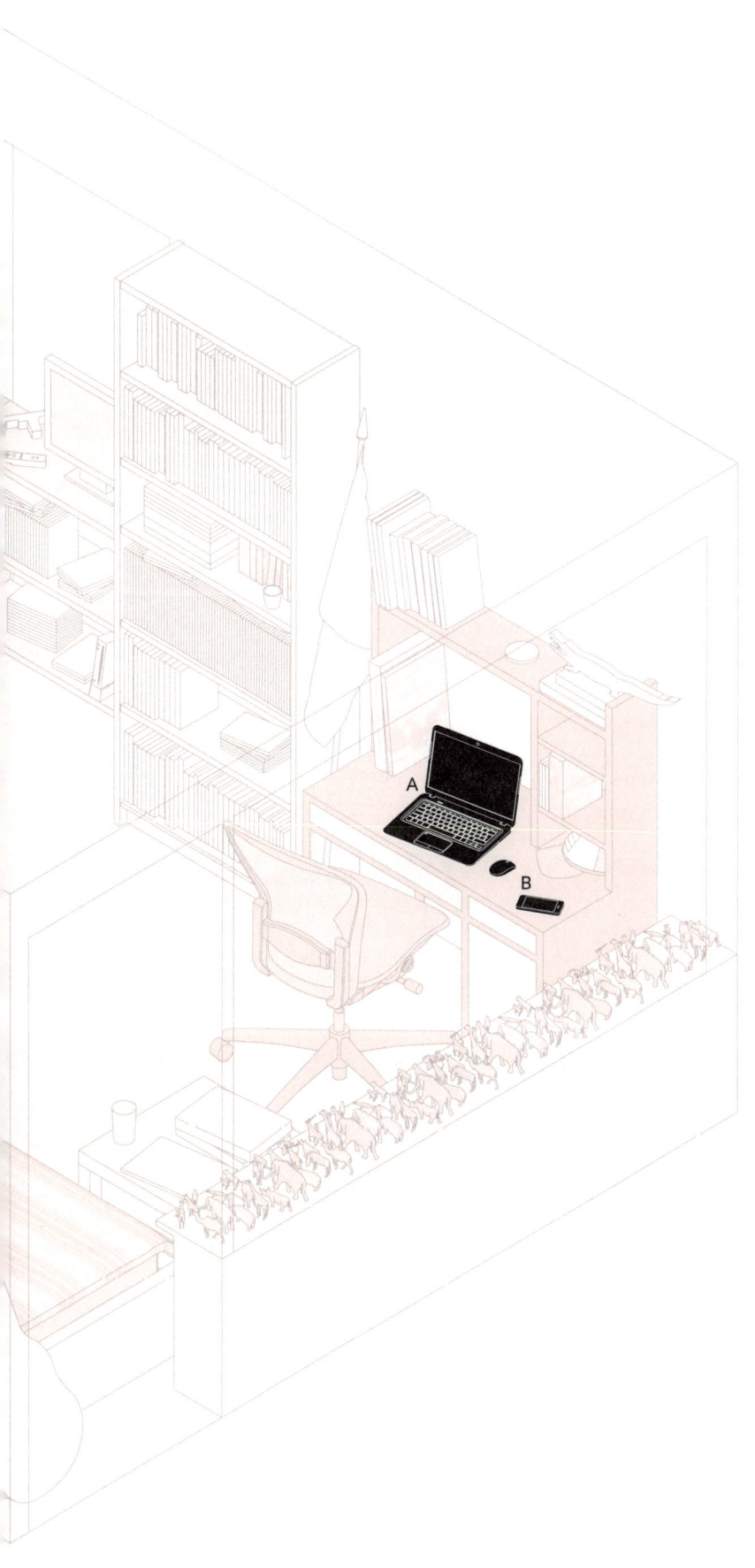

A   Packard Bell Easynote TJ-65
B   Samsung Galaxy S4

6 Conclusion: A Post-Representational Approach to Graphic Design

Blind Maps and Blue Dots

In this concluding chapter I will reflect on the outcomes of my research, and formulate an answer to the research question: "What can a post-representational reading of contemporary mapmaking practices reveal about the blurring of the producer–user divide in graphic design?" First, I will propose a new approach to graphic design. By focusing on tools, the understanding of the field can be expanded with a variety of practices of production and use. I will then present a new conceptual model, a timeline of technological thresholds, to better understand the transformation of the field of graphic design. Using the insights gained from this new approach, I will connect the design field and cartography and explain why mapmaking is a relevant field to study the transformation of graphic design. From case studies of three contemporary mapmaking practices I have developed two important visual concepts, the Blind Map and the Blue Dot, that, together with the various ways they interact with each other, constitute a post-representational approach to graphic design. I will present the ambiguous strategies I developed in my graphic design practice as a response to this post-representational approach. I will conclude by addressing the need for alternative and additional languages in multidisciplinary discourse, in research in general, and in artistic research in particular.

Transformation of Graphic Design

As stated earlier, ever since its inception in the late nineteenth century, graphic design has been closely tied to industrial production. The field originated as a specialized activity concerned with creating, editing and reproducing visual information. The shift towards digital modes of production has transformed graphic design in a fundamental way. The desktop publishing revolution of the 1980s combined several specialist activities of the industrial production of visual information into a single machine that is accessible to all: the computer and its software. This universal tool, the interactive possibilities of digital media, and the easy exchange of visual information through networks like the Internet, led to a different relation between the creators, editors, producers and users of visual information. In my view, this has happened to the extent that a clear distinction between the various parties in the information chain no longer exists.

The situation in the field of graphic design in the early 2000s has been characterized by American author, curator and graphic designer Andrew Blauvelt as suffering from a malaise of disciplinary formlessness.[1] The field had expanded way beyond its roots in print, and while other parties—read: nondesigners—were using the tools of graphic design, those with a specific training or specialized professional practice in graphic design on their part were appropriating the roles of authors, archivists, artists, curators, editors, educators, entrepreneurs, programmers, publishers, researchers, storytellers, tool makers, visual journalists and others.

While Blauvelt describes graphic design as a field that lacks coherence and is totally dispersed, I propose to take his assessment one step further, by claiming that the democratization of the means of production and the diversification of its output are signs of the discipline dissolving and becoming part of a larger field. This larger field is focused on graphic representation. It encompasses many

1   Blauvelt, "Towards Critical Autonomy, or Can Graphic Design Save Itself?," 8.

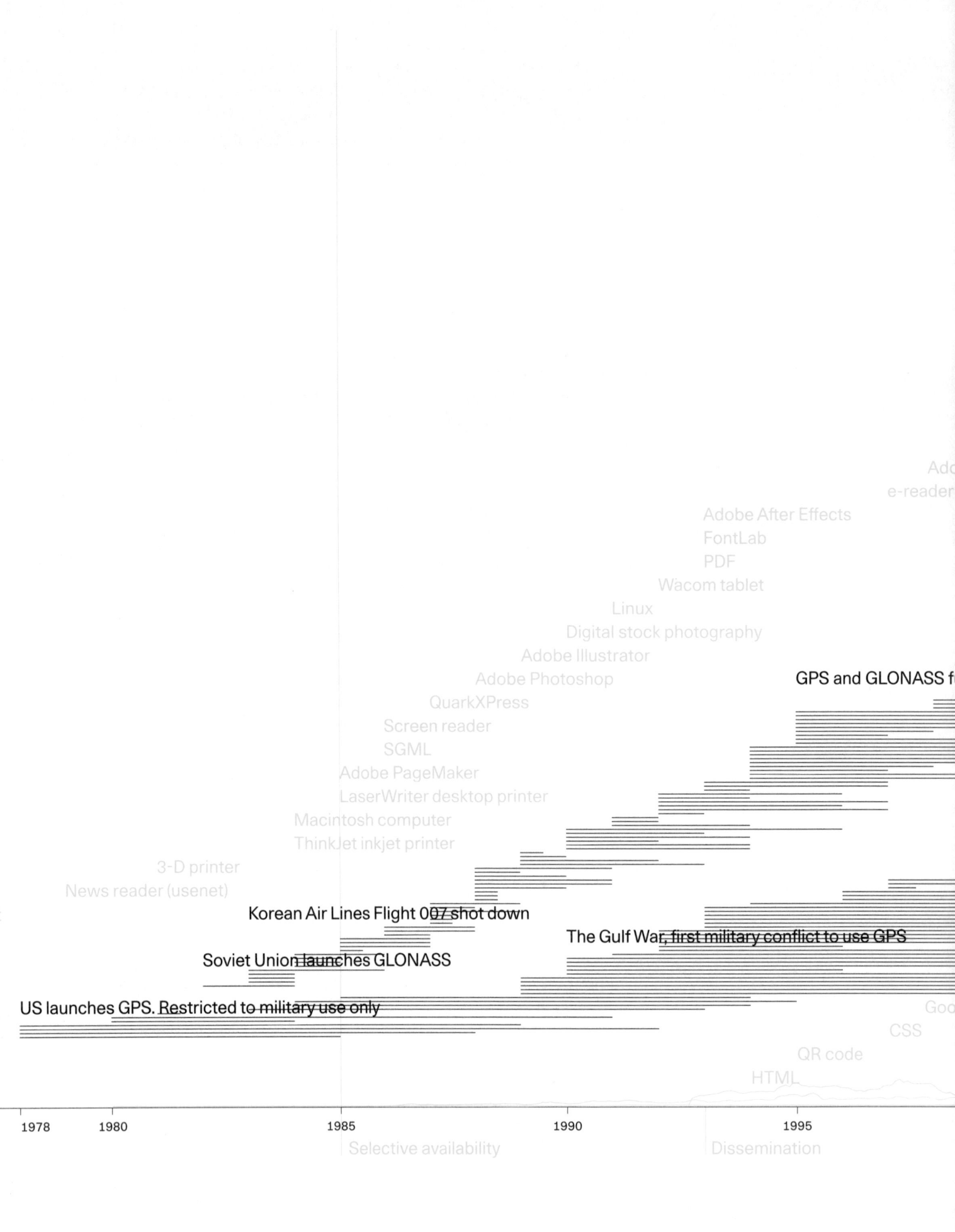

Adc
e-reader

Adobe After Effects
FontLab
PDF
Wacom tablet
Linux
Digital stock photography
Adobe Illustrator
Adobe Photoshop
QuarkXPress
Screen reader
SGML
Adobe PageMaker
LaserWriter desktop printer
Macintosh computer
ThinkJet inkjet printer
3-D printer
News reader (usenet)

GPS and GLONASS f

ft

Korean Air Lines Flight 007 shot down

The Gulf War, first military conflict to use GPS

Soviet Union launches GLONASS

US launches GPS. Restricted to military use only

Goo
CSS
QR code
HTML

1978   1980   1985   1990   1995

Selective availability   Dissemination

In this concluding series of visualizations, the most important technologies
and tech companies from this book are shown over time, compared with one
another and the technological thresholds model.

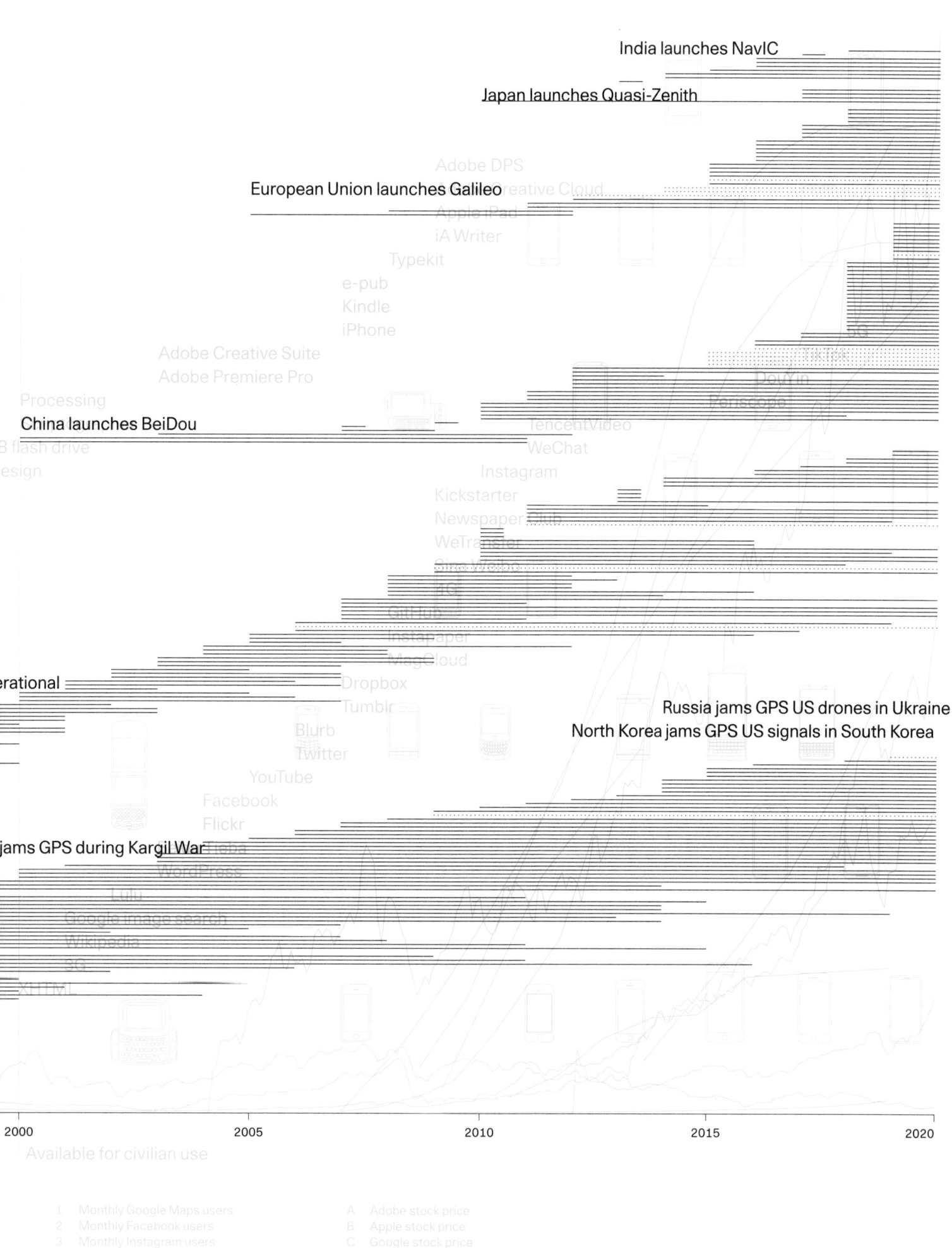

India launches NavIC

Japan launches Quasi-Zenith

Adobe DPS
European Union launches Galileo Creative Cloud
Apple iPad
iA Writer
Typekit
e-pub
Kindle
iPhone

Adobe Creative Suite
Adobe Premiere Pro
Processing
China launches BeiDou
B flash drive
Design

Adobe DPS

5G

Douyin
Periscope
TencentVideo
WeChat
Instagram
Kickstarter
Newspaper Club
WeTransfer
Sina Weibo
4G
GitHub
Instapaper
MagCloud

Dropbox
Tumblr
Blurb
Twitter
YouTube
Facebook
Flickr
jams GPS during Kargil War Tieba
WordPress
Lulu
Google image search
Wikipedia
3G
XHTML

erational

Russia jams GPS US drones in Ukraine
North Korea jams GPS US signals in South Korea

| 2000 | 2005 | 2010 | 2015 | 2020 |

Available for civilian use

distinct practices that are rooted in a variety of disciplines, but that use similar tools to create, record, edit, produce and distribute visual information. Practitioners with academic, creative, journalistic and other backgrounds use the same hardware and software and are dealing with similar concerns: What do I present, how do I show it and in what way do I make it public?

As I see it, Blauvelt's evaluation of graphic design's recent developments is too much focused on what happened to the persona of the graphic designer. In his writings, the designer in the digital age emerges as someone who has been struck by friendly fire.[2] I use this very term that has the connotation of injustice and fatality because I detect a slight sense of despair, primarily about the role of the designer, in Blauvelt's analysis of the current state of graphic design. The specialist tools that gave the graphic designer increased control and power in the digital age, now threaten to make her obsolete. In my view, when evaluating graphic design, especially as it is morphing into the larger field of graphic representation, we should look beyond the designer.

I propose to reconsider graphic design by looking at the technologies that gave rise to the field and that are now transforming the field into something else. That is not to say that technological developments are the only explanation for the current state of graphic design. For example, revised concepts about authorship and an experimental attitude towards exploring and breaking the rules of the graphic design craft influenced the field in the last decades of the twentieth century. However, these evolving ideas are centered exclusively on the designer, and do not consider the changing position of the user.

I have developed a model for understanding the developments in the field of graphic design. This is a chronological presentation of various graphic technologies that enables the situating of various practices of production and use. My model is a timeline of technological thresholds, that is, an overview of the technologies for creating, recording, editing, producing, distributing and accessing visual information. In this timeline, I have identified three sets of tools related to mechanization, digitization and dissemination. Mechanization refers to the technologies of the industrial production of graphic information that enabled the graphic designer to emerge as a specialist in the production of visual information. Digitization of production tools has impacted the role of the designer, the tasks she performs and her role in the information chain. Dissemination technologies impacted the speed and expanse of the distribution of information, and the interaction and exchange of that information with others. Although different in nature, the two latter sets of tools reinforce each other in what I believe to be a significant transformation of the graphic design field.

New questions arise when the transformation of graphic design is researched on the basis of the tools it uses. What, for instance, is the impact on the field of practitioners with roots in other disciplines, but whose practices are virtually the same as graphic design, because they use the same tools? Take cartography, for instance. Graphic design and mapmaking have different origins and concerns, but their contemporary practices have much in common. Both use similar tools and the digitization of those tools has enabled new players to enter the fields. The impact of technological developments on cartography seems greater and clearer than on graphic design. In my view, cartography, both the activity of mapmaking and the theory of maps, has had a head start on graphic design in dealing

160

with the impact of digital technologies. Therefore I consider it a testing ground to understand the transformations of graphic design. For my research, I selected three mapmaking practices that I regard as Petri dishes to survey, analyze and test the transformation of the field of graphic design.

Post-Representational Cartography

For the investigation of three contemporary mapmaking practices I adopted notions from post-representational cartography. This theoretical approach regards a map as a process rather than a fixed spatial representation. Seen as emergent, a map is never finished; it is in a constant state of becoming. Producing a map is a continuous process, and so is its use. Therefore, the binary divisions between production and use and between producer and user are no longer valid.

What I take from post-representational cartography is the consideration that "making" and "using" are not consecutive processes but parallel tracks. Furthermore, post-representational cartography provides ways to critically think about a variety of practices engaged with cartography beyond those of professional specialists. I also appreciate the shift of focus from the end product to how the map works, how it shapes our understanding of the world, and how it codes that world.

I disagree with the theory of post-representational cartography in that I put as much emphasis on the processual nature of the map as on the ever-changing position of the user-producer. For me, both the map and the mapmaker/user have a continuous dynamic status.

Another point where I deviate from post-representational cartography is that I see differences in the extent to which a map can be processual. Certain maps offer the user a bigger role by literally putting her on the map, such as in Google Maps where the Blue Dot highlights the user's position. In the case of digital maps, where movements or locations are recorded, the user can play a greater role as a co-producer than with traditional maps on paper. However, there is also a shadow side to this digital function in which the data generated by the movements of a user increases the power position of the producer.

It is not only a matter of functionality whether a map can be seen as processual. The method of distribution, and whether the map is part of a conversation in which sources and design choices can be queried and subsequently updated, also contribute to the extent to which a map is processual.

There is a third factor that determines the extent to which a map is processual and that is its design. The method of presentation—the way it looks, the amount of information it contains, the legend that has been used, et cetera—facilitates the process by which the user engages with, and, as it were, "enters" the map.

A Post-Representational Reading of Three Contemporary Mapmaking Practices

In this book I have adopted a post-representational approach to three case studies of contemporary mapmaking. Based on my fascination for the processual aspects of cartography, I identified three mapmaking practices for further study: the Blue

2   *Friendly fire* is a military term for an accidental attack by a force on its own army or allied troops.

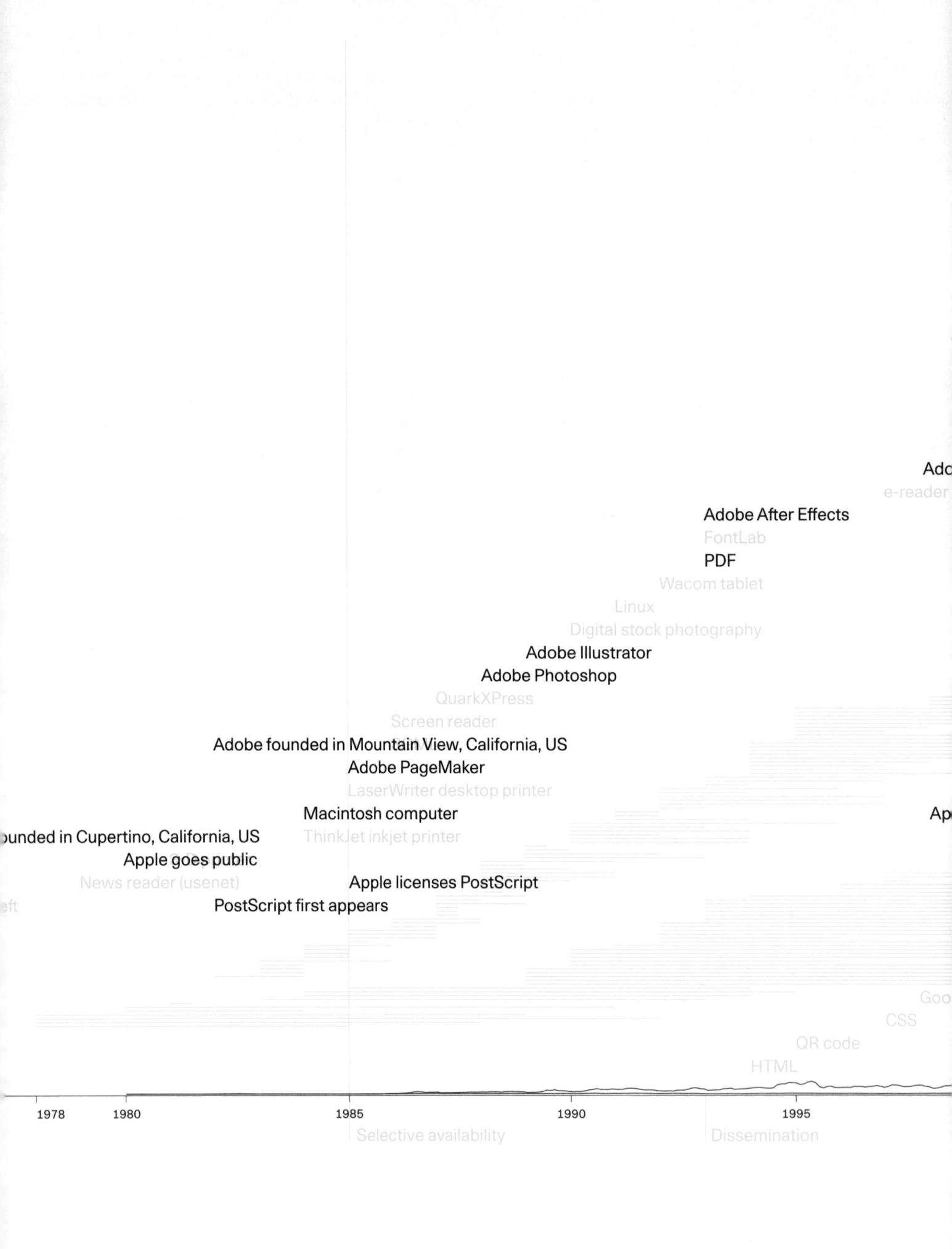

Adc

e-reader

Adobe After Effects

FontLab

PDF

Wacom tablet

Linux

Digital stock photography

Adobe Illustrator

Adobe Photoshop

QuarkXPress

Screen reader

Adobe founded in Mountain View, California, US

Adobe PageMaker

LaserWriter desktop printer

Macintosh computer

ThinkJet inkjet printer

Ap

ounded in Cupertino, California, US

Apple goes public

News reader (usenet)

ft

Apple licenses PostScript

PostScript first appears

Goo

CSS

QR code

HTML

1978   1980   1985   1990   1995

Selective availability   Dissemination

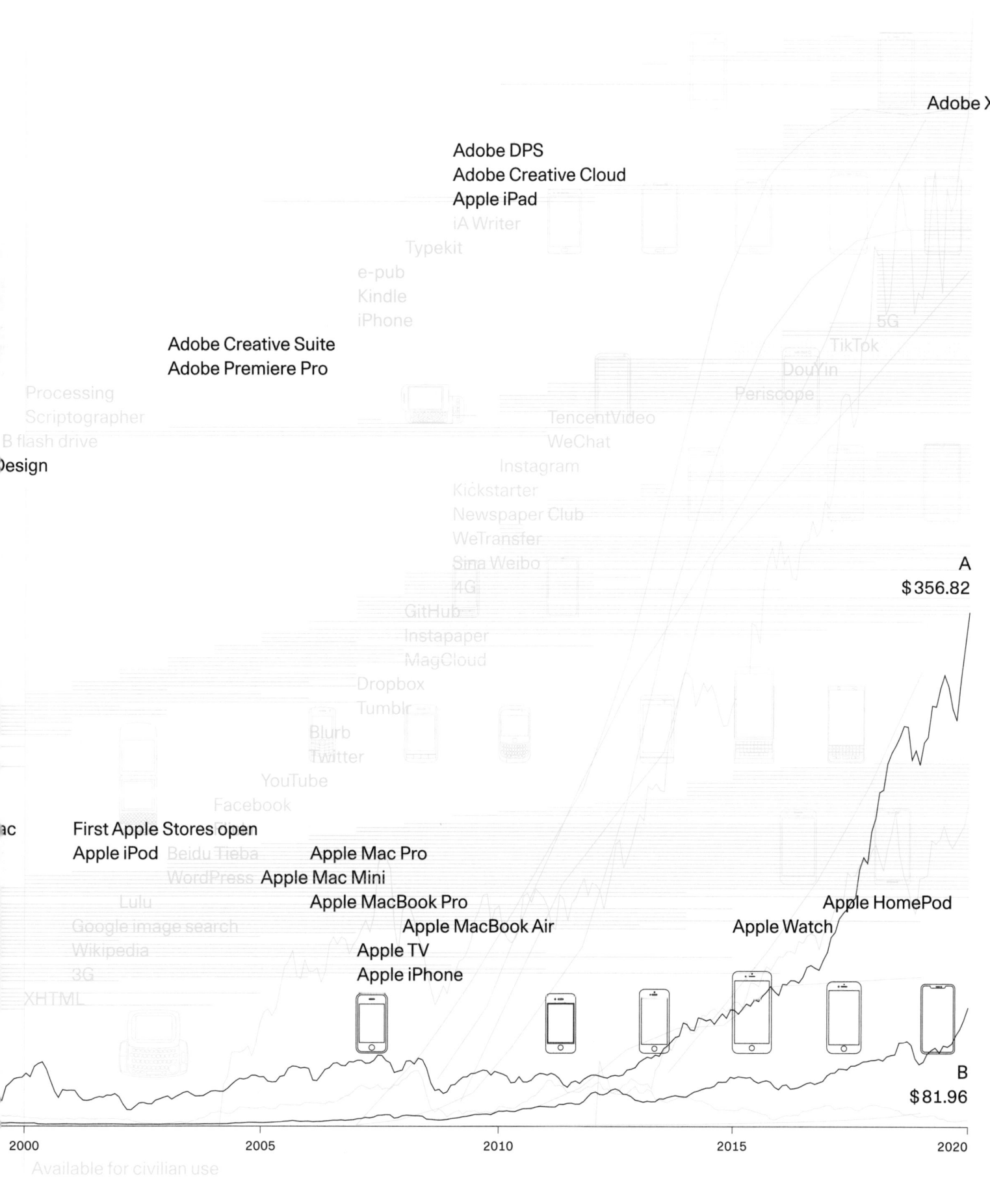

Adobe XD

Adobe DPS
Adobe Creative Cloud
Apple iPad
iA Writer
Typekit
e-pub
Kindle
iPhone

5G
TikTok
DouYin
Periscope

Adobe Creative Suite
Adobe Premiere Pro

Processing
Scriptographer
B flash drive
Design

TencentVideo
WeChat
Instagram
Kickstarter
Newspaper Club
WeTransfer
Sina Weibo
4G
GitHub
Instapaper
MagCloud
Dropbox
Tumblr
Blurb
Twitter
YouTube
Facebook

A
$ 356.82

ac
First Apple Stores open
Apple iPod   Beidu Tieba   Apple Mac Pro
                    WordPress   Apple Mac Mini
Lulu   Apple MacBook Pro
Google image search   Apple MacBook Air
Wikipedia   Apple TV
3G   Apple iPhone
XHTML

Apple HomePod
Apple Watch

B
$ 81.96

2000   2005   2010   2015   2020

Available for civilian use

1   Monthly Google Maps users   A   Adobe stock price
2   Monthly Facebook users   B   Apple stock price
3   Monthly Instagram users   C   Google stock price
4   Monthly Twitter users   D   Facebook stock price
5   Strava members   E   Twitter stock price
6   Worldwide sales of smartphones
7   Worldwide sales of smartphones running   Operational satellites
    on Android   Satellites in reserve / Testing

Dot, the location function in Google Maps; the Strava Global Heatmap, a world map showing the activities of users of the fitness app Strava; and the "situation in Syria maps," a regularly updated map of the Syrian conflict produced by an Amsterdam teenager. Together, these practices deal with a variety of aspects that are important in graphic representation today: the user as co-creator and the relation between a functional and critical understanding of graphic products (the Blue Dot); the motives behind, and the effects of, certain visual strategies and the understanding and misunderstanding of certain typologies (the Strava Global Heatmap); the specialist–amateur divide and the role of expertise in graphic production (the situation in Syria).

In the Blue Dot case study I used the three different modes of cartographic thinking presented above to investigate Google Maps, which is the most used map today. Each of these modes of thinking generates considerations that become criteria for evaluating the look and functionalities of a map. Conversely, how the design of a map is perceived will be informed by the viewer's conceptual understanding of the cartographic product. From this it follows that a purely functional reading of the map, without its being questioned on a conceptual level, does not suffice. Nor does a critical reading of the map solely focused on understanding the hidden structures suffice, because it does not provide criteria for how a map should function. Building on this idea, it can be stated that design is a way to invite theory. In mapmaking, theory and design are intertwined. A theoretical understanding of a product is necessary to perceive the significance of its design.

Google Maps is the quintessential example of a processual map. It has a fundamentally emergent status: it looks empty, and its colors are pale as if the map anticipates being filled with a highlighted location or route. Opening the app, the software does not show a map that is complete but one that is the starting point of a process of searching, scrolling and zooming in and out. The My Location function in Google Maps enables users to perceive their position as a small blue dot. This Blue Dot puts the user on the map, literally. Besides being a visual sign, it is also a symbol, characterizing a next stage in thinking about production and use. Here, the binary division between producer and user no longer applies. The user is not the recipient at the end of a process, but to a certain extent she is the co-creator of a graphic product. At the same time the Google Maps user is a victim, as Google exploits her data as raw material. On another level she is prey, because orientation by way of using Google Maps is not a process of comparing a site with its diagrammatic representation, but about seeing a single version of that reality. One that is blind to any version but that of Google.

The second case study of the Strava Global Heatmap, a map based on data generated by users of the fitness tracking app Strava, considers knowledge from a variety of fields in order to investigate the technological, economic, social and cultural aspects of the Global Heatmap. I argue that there is a disconnection between the data that form the basis of the Strava Global Heatmap and the information it displays. The map uses a deceivingly familiar formal language, but in fact it is a type of visualization that we are not yet familiar with. The map's content is manipulated, using a visual language with similarities to photography. If defamiliarization is the technique of disrupting the user's expectations to stimulate fresh perceptions, then the visual strategies at work in the Global Heatmap do exactly the opposite. The map's design renders it in a soothing, familiar format that prevents users from questioning it. The deceiving familiarity of the Strava Global Heatmap raises an intriguing question: Can a map that is misread still be called a map? This issue becomes especially relevant when a map is considered to be a process, an unfinished product that needs a user to complete it.

The case study of the Strava Global Heatmap demonstrates how strategies of visual deception, caused by familiar references, are at work on several levels. For example, the platforms offered by social media companies such as Strava are presented as public spaces, whereas in fact they are private spaces designed to keep users there, maximizing the number of moments to monetize their presence. Another example is the Global Positioning System (GPS), the technology essential in all three case studies. Originally, GPS was a defense technology, which was later opened up to empower civilians, but at the same time the move was a shift in power from military to economic dominance. GPS tracks are personal expressions of users exploited by technology companies in their pursuit of total vision and economic gain.

The third case study, of the "situation in Syria maps," examines the practices of amateurs who create maps of armed conflicts. I consider their work to be an anti-representational form of cartography. One of the aims of representational cartography is to improve the effectiveness of a map, to carefully balance the content of the map and its graphic container. Amateur conflict mapmakers lack the skills and knowledge to make the kind of sophisticated map that representational cartography pursues. Given the nature of their subject matter, and the (predominantly online) debates they are part of, amateur conflict mapmakers put more emphasis on the process of unearthing data, on showing what otherwise would be hidden, than on the representation of information, on mapmaking. Their maps are visibilizations rather than visualizations.

The maps of amateur conflict mapmakers may lack certain visual qualities, but they may have other merits when considered processually. Conflict maps have a high level of accountability because they are part of a public debate. The maps by amateur conflict mapmakers are shared on online platforms, where they are shown with their sources, together with previous versions, and surrounded by discussions about claims made on the map. In my analysis of the work of Thomas van Linge, an Amsterdam teenager who mapped the conflict in Syria, I have shown that a map can be embedded in several debates. These can vary from open dialogues on social media to more internalized debates that show the process and considerations of the mapmaker when comparing several maps. These debates highlight the processual nature of conflict maps. They are never finished, constantly updated, and open to discussion.

Nonspecialists mapmakers mostly use free and generic software to produce maps. Specialist designers, on the other hand, mainly use commercial specialist tools. Tools shape the practices that use them. Specialized software generates a specialized practice because of the economic model employed by software manufacturers, the demands of an industry that imposes specific exchange formats, and the curricula of educational institutes training new practitioners. By means of improvisation, and the use of generic tools, amateurs create different kinds of graphics editing practices that are not based on specialization.

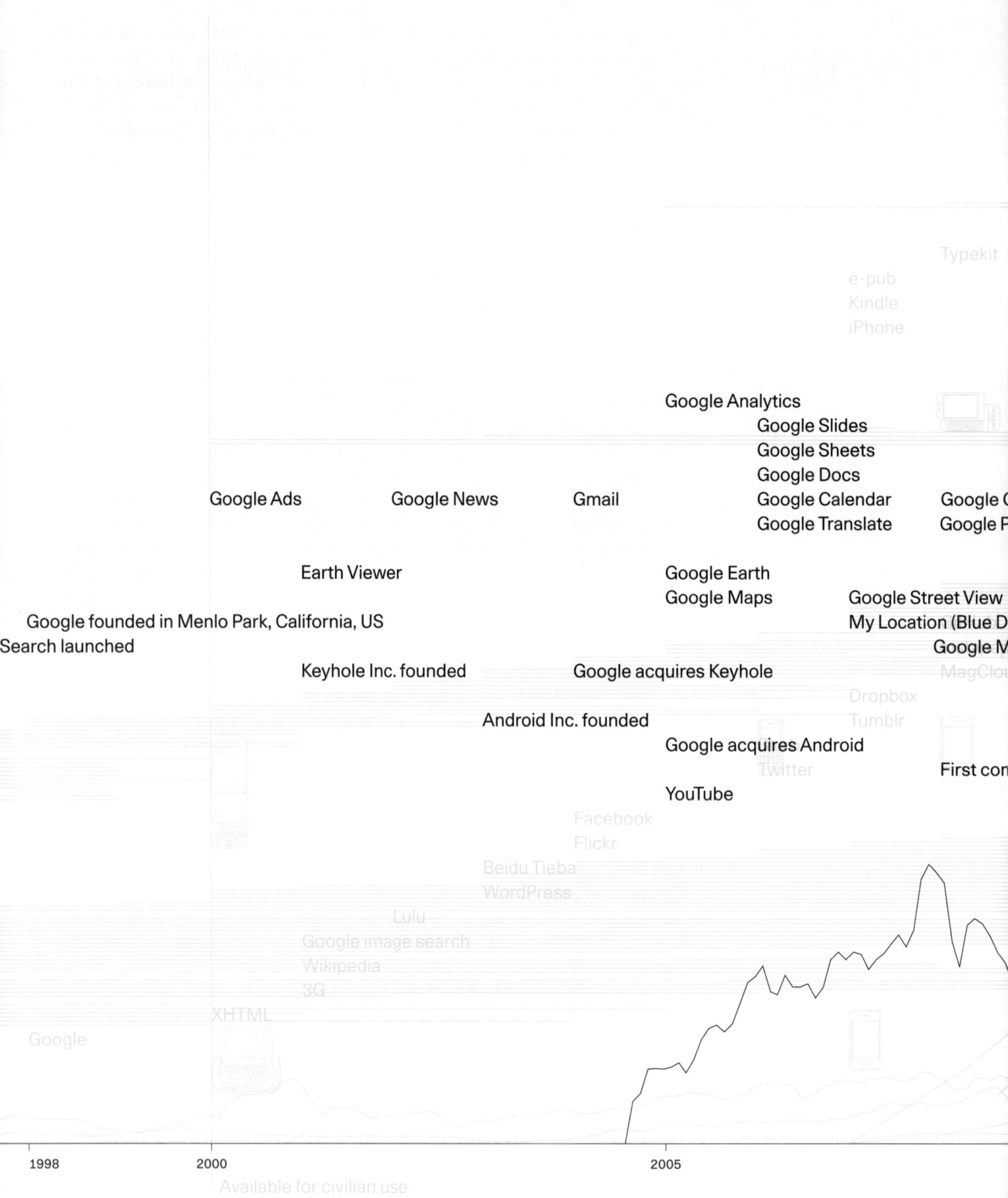

Typekit
e-pub
Kindle
iPhone

Google Analytics
Google Slides
Google Sheets
Google Docs
Google Ads   Google News   Gmail   Google Calendar   Google C
Google Translate   Google F

Earth Viewer   Google Earth
                                          Google Maps   Google Street View
Google founded in Menlo Park, California, US   My Location (Blue D
Search launched   Google M
                    Keyhole Inc. founded   Google acquires Keyhole   MagClou

                                                                        Dropbox
                    Android Inc. founded   Tumblr
                                          Google acquires Android
                                          Twitter   First con
                    YouTube

                    Facebook
                    Flickr
                    Beidu Tieba
                    WordPress
                    Lulu
                    Google image search
                    Wikipedia
                    3G
                    XHTML

Google

1998   2000   2005

Available for civilian use

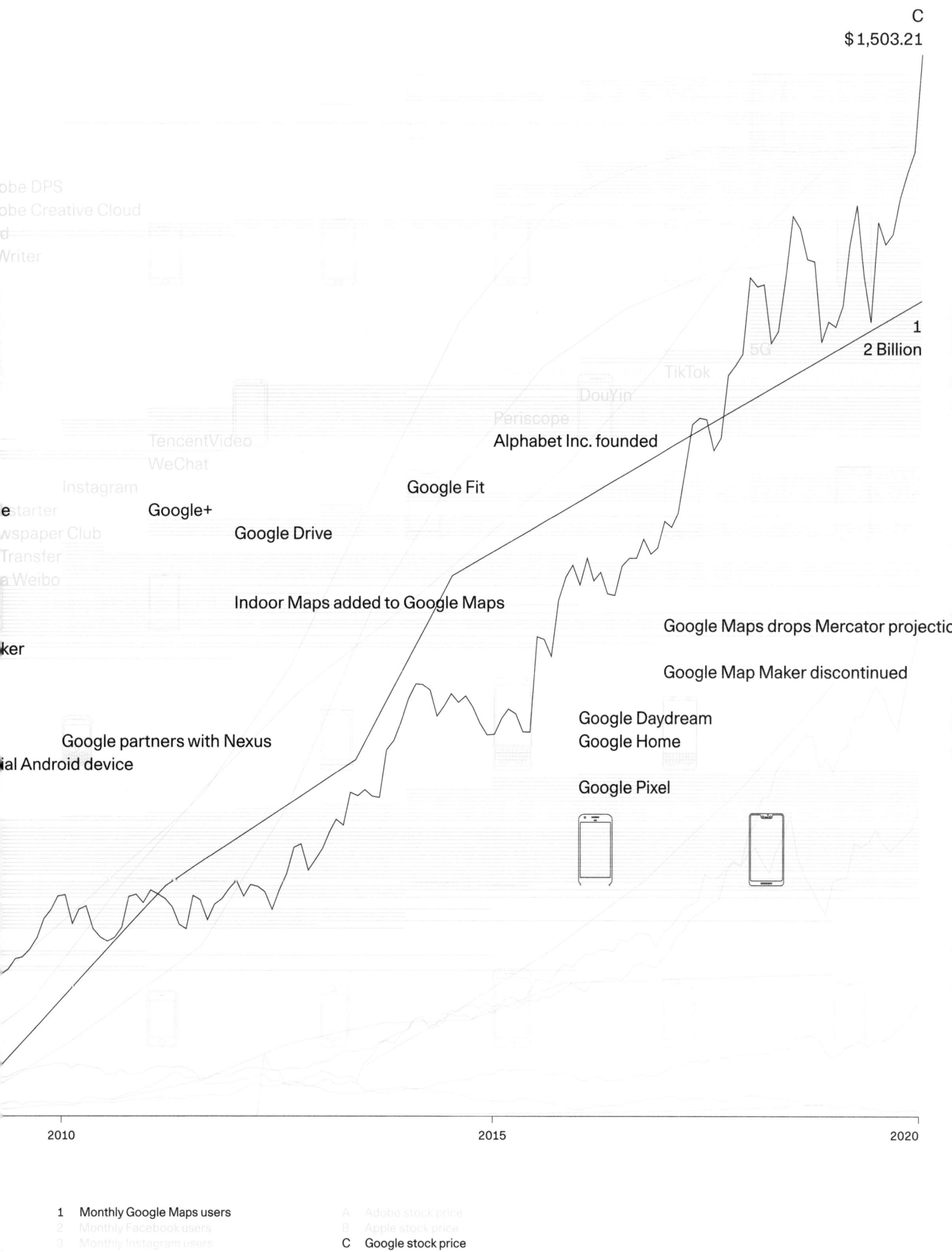

C

$ 1,503.21

obe DPS
obe Creative Cloud
d
Writer

5G

TikTok

DouYin

Periscope

Alphabet Inc. founded

TencentVideo
WeChat

Google Fit

Instagram

e starter
wspaper Club
Transfer
a Weibo

Google+

Google Drive

1

2 Billion

Indoor Maps added to Google Maps

Google Maps drops Mercator projectio

ker

Google Map Maker discontinued

Google Daydream
Google Home

Google partners with Nexus

ial Android device

Google Pixel

2010   2015   2020

1   Monthly Google Maps users   A   Adobe stock price
2   Monthly Facebook users   B   Apple stock price
3   Monthly Instagram users   C   Google stock price
4   Monthly Twitter users   D   Facebook stock price
5   Strava members   E   Twitter stock price
6   Worldwide sales of smartphones
7   Worldwide sales of smartphones running   Operational satellites
     on Android   Satellites in reserve / Testing

A Post-Representational Approach to Graphic Design

Theoretical research into post-representational cartography and insights from the case studies enabled me to formulate two visual concepts: the Blind Map and the Blue Dot. Both deal with the fundamental processual and dynamic status of respectively the map and its producer-user. They are transdisciplinary and applicable to other forms of graphic representation. The two visual concepts, and the various ways they interact, together constitute a post-representational approach to graphic design.

The Blind Map is based on a type of map that does not contain any text labels and that is mainly used in education to let pupils fill in the missing names in tests. Graphic products are Blind Maps, they are in a permanent state of emergence, never fully formed, but completed every time a user engages with them. Google Maps is the quintessential example of a Blind Map: its colors are pale as if the map anticipates being filled with a highlighted location or route, and its crop or zoom to be changed. In essence, Google Maps is a processual map. For a graphic product to be a Blind Map it should offer opportunity for a user to engage with it. In my experience as a designer, one way to do this is to create space, moments of emptiness, for a user to project her ideas and actions on, to fill in, to enter the graphic product.

The Blue Dot is both a visual sign indicating the user's location on a digital map and an emblem marking a different phase in the thinking about production and use of maps in which the binary division between producer and user no longer applies. The Blue Dot is about control and being controlled. The user takes control by producing a map based on her location, choice of zoom level and preference for additional layers of information; she is as much a producer as a user of the graphic product. The extent to which the user can also be a co-producer of the map is limited by what is permitted by the producer and publisher of the map and to a lesser extent by possibilities offered by the software and the device on which the map is accessed. At the same time, the user is being observed by the publisher of the map that is exploiting the generated user data. In that regard the producer is also a user. This fluctuating dynamic between producing and using can also be observed in other graphic products than maps: designers developing graphic tools like apps or typefaces for others to work with, designers developing co-creation projects through online platforms or amateurs producing photo books of holiday snaps using Internet-based software. These are all examples of practices where the producer–user divide dissolves.

The two concepts I introduced above are interrelated. The Blue Dot gives the user a presence in the process of graphic representation, but in order to take on this role the user needs to be given room and agency to co-produce. The Blind Map offers space for the user to take this position. The use of a graphic product is an encounter between two dynamic entities, the ever-emergent product and its producer-user, who is constantly fluctuating between making and using. The insights I gained during my research are a consequence of this encounter between two dynamic entities.

If the graphic product is regarded as constant, and the information chain of publisher, editor, producer, user as solid, then the relation between the product and each participant in the information chain is distinct but fixed. However, if we regard both the graphic product and the participants in the information chain as entities in flux, then their encounter is ever-changing, causing various misalignments. The dynamics of varying oblique positions of formats, users and producers result in misunderstandings and false expectations of the information contained in a graphic product, or of the presumed knowledge and authority of a participant in the exchange.

Examples of misalignments are the deceiving familiarity found in the various case studies. For instance, the light colors in Google Maps make it look like a paler version of something well-known, the memory of a map. Or the Strava Global Heatmap, which uses visual strategies that are reminiscent of long-exposure photography that create an image type the user has seen before and thinks she knows how to read. Or the use of design tropes of traditional cartography, which makes the maps of Thomas van Linge appear neutral and authoritative, and provide the mapmaker with the status of a specialist.

An ever-changing graphic product gains authority when it is embedded in a public debate, where information about the reasoning behind decisions is provided, where sources can be checked, and where users and producers can engage in a conversation. Besides featuring in this public debate, a graphic product may also conduct an internal debate, which can be observed when various stages of the product are published, revealing the ongoing process of production and reproduction. Another example of this internal debate are the traces of previous steps that are incorporated in later editions of the graphic product.

The publishing strategies of amateur conflict mapmakers on social media platforms are an example of how these mapmakers engage in a public debate. Another is Strava's publication of a background article on the making of the Global Heatmap. An opposite case is Google, who does not provide a map key of Google Maps, which causes me as user to be suspicious of the map and the motivation of the mapmaker.

In the dynamic constellation of production and use, devices, digital platforms and tools play a pivotal role as the mediators between the graphic product and its producer-user. The technologies that enable creation, recording, editing, production, distribution and access were always positioned between the product and the producer-user. But because the status of the graphic product and that of the participants in the information chain is dynamic, the mediating technologies between them gain importance. Take, for instance, a map displayed on a smartphone. It has no predefined crop, but is based on the location of the user. The proportions, resolution and size of the screen are meaningful elements in the design of the map, even if this is only partly controlled by the mapmaker. Another example of a mediating technology that plays a pivotal role in the exchange of producer and user is design software that has incorporated various design decisions, like typographic settings, color palettes and predefined sizes. One could wonder in such a case to what extent the design is solely made by the graphic designer?

The democratization of design tools and the widespread digital networks for exchange, such as the Internet, have ensured easy access to the means of production for anyone who has access to a computer, smartphone, or other digital

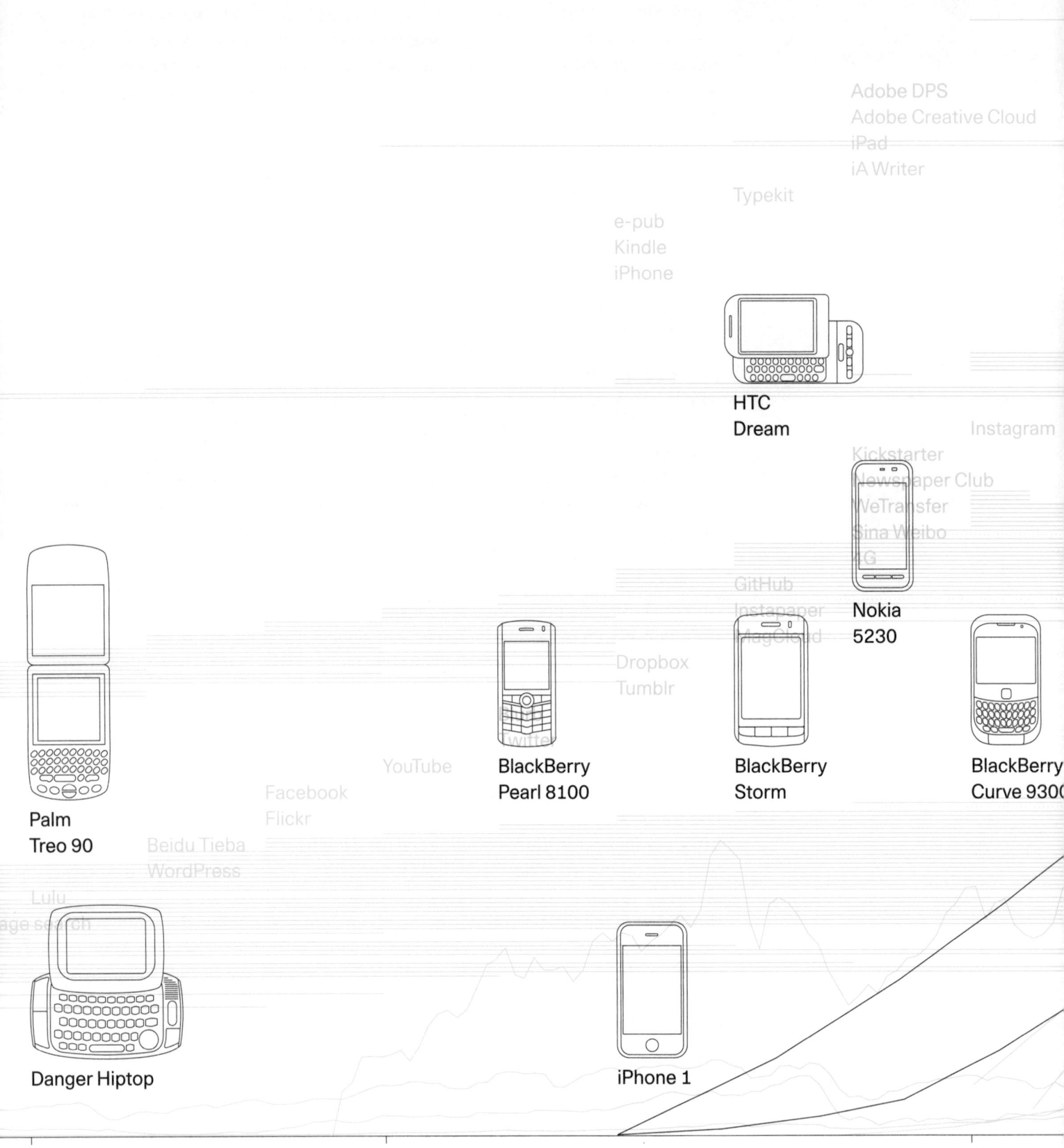

Adobe DPS
Adobe Creative Cloud
iPad
iA Writer

Typekit

e-pub
Kindle
iPhone

HTC
Dream

Instagram

Kickstarter
Newspaper Club
WeTransfer
Sina Weibo
4G

GitHub
Instapaper
MagCloud

Nokia
5230

Dropbox
Tumblr

Twitter

BlackBerry
Pearl 8100

BlackBerry
Storm

BlackBerry
Curve 9300

YouTube

Facebook
Flickr

Beidu Tieba
WordPress

Lulu
Image search

Palm
Treo 90

Danger Hiptop

iPhone 1

2002

2005

2010

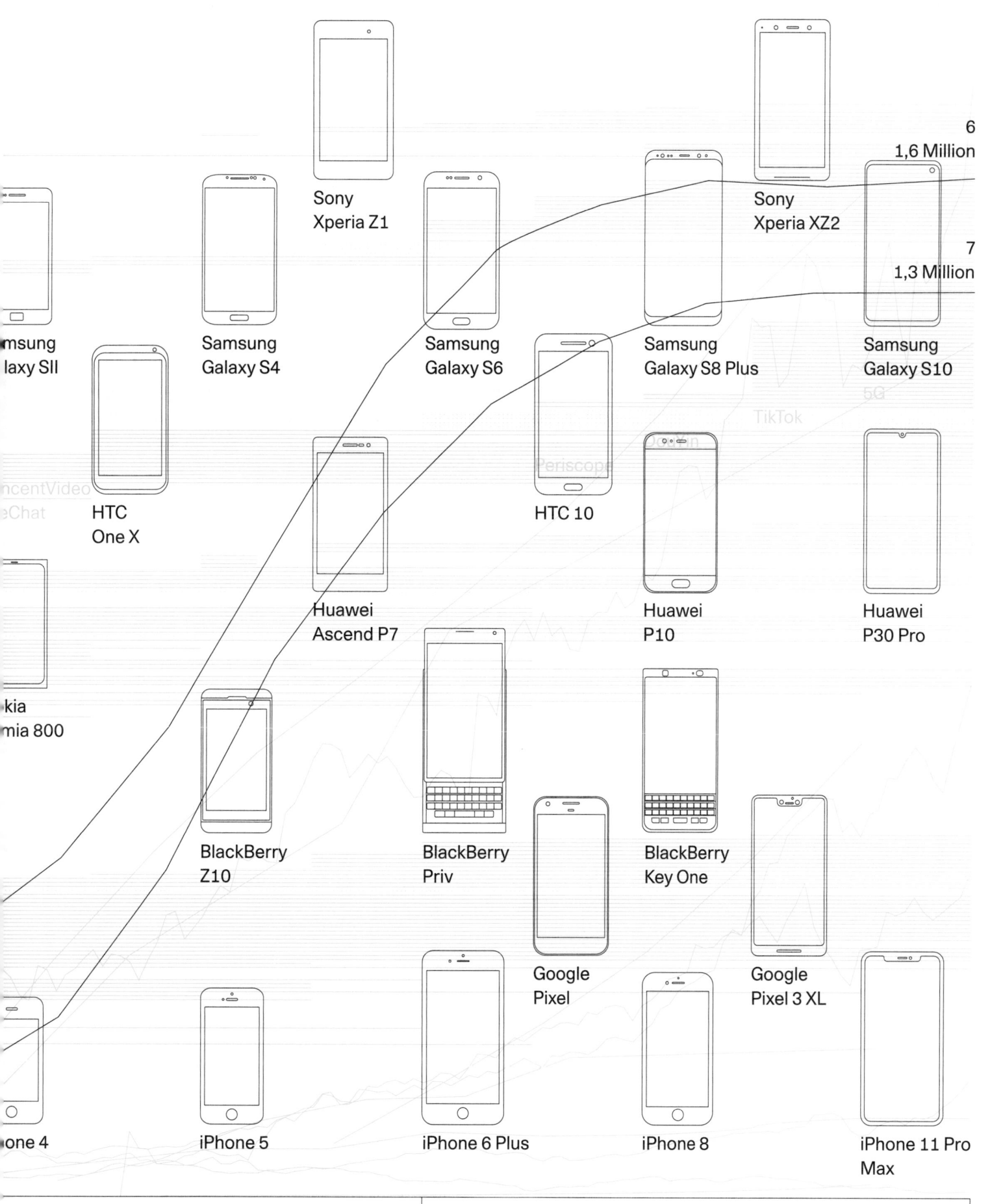

Samsung Galaxy SII

HTC One X

Samsung Galaxy S4

Sony Xperia Z1

Samsung Galaxy S6

Samsung Galaxy S8 Plus

Sony Xperia XZ2

Samsung Galaxy S10

6
1,6 Million

7
1,3 Million

Huawei Ascend P7

HTC 10

Huawei P10

Huawei P30 Pro

...kia ...mia 800

BlackBerry Z10

BlackBerry Priv

BlackBerry Key One

Google Pixel

Google Pixel 3 XL

...one 4

iPhone 5

iPhone 6 Plus

iPhone 8

iPhone 11 Pro Max

2015

2020

1 Monthly Google Maps users
2 Monthly Facebook users
3 Monthly Instagram users
4 Monthly Twitter users
5 Strava members
6 Worldwide sales of smartphones
7 Worldwide sales of smartphones running
   on Android

A Adobe stock price
B Apple stock price
C Google stock price
D Facebook stock price
E Twitter stock price

Operational satellites
Satellites in reserve / Testing

device. Tools incorporating design knowledge and the decreasing expertise of users who lack training but do have access to the tools are opposite phenomena that reinforce each other. These mutually opposing movements increase the enormous impact of technology companies on the graphic design field. In fact, the tech industry partly determines what is produced because its tools incorporate design knowledge in the menus and settings. These companies already have a dominant position because they develop the file formats for exchange and production and thus dictate how visual information is produced. Moreover, the products of software and hardware companies have a strong presence in the curricula of design schools, so that they also determine how production will take place in the future. The dominance of software companies in the design field manifests itself in the economic model used to exploit their products that almost exclusively leaves room for highly specialized practices. In this way, technology companies not only determine what is made, how it is produced, now and in the future, but also how designers work.

The notion of the incredibly powerful position of technology companies extends beyond the realm of the design field to the full spectrum of our lives. They monopolize and centralize the technologies that allow us to find our way, record our activities and ideas, and share these in our conversations with others. The data we generate, in turn, becomes the property of these powerful companies who then exploit it as raw materials to increase their economic power. The notion of friendly fire also seems to apply to this catch-22 situation: we are casualties of the tools that we can no longer do without.

Graphic designers and other editors of visual information know how to escape this seemingly inevitable model of production and specialization, by employing insurgency tactics: using generic tools or creating their own means of production. Although certain designers do exactly this, it is not yet a general praxis. In my practice mainly tools from the large commercial technology companies are used. With the insights from this research I will certainly investigate how we can develop more tools of our own. This may be even more urgent in the context of education. What is the significance of the dependence on software companies for the reevaluation of educational curricula? And: the fewer designers, and people in general, who actually depend on tools developed by others, the less likely they are to be struck by friendly fire.

Ambiguous Strategies

In my design practice I have developed strategies of incorporating ambiguity as an antidote to the misalignments in the encounters between the producer-user and the graphic product. Following the ideas of American visual theorist and culture critic Johanna Drucker, I acknowledge the fundamentally constructed nature of data. By incorporating the ambiguity and uncertainty of data in the design of information, I highlight and challenge the manipulations in visualizations. This can be done by making maps with more nuanced legends that challenge the assumptions implied by the use of certain colors, shapes, lines and transitions. By using a variety of visualization typologies rather than one singular type, so that the selective nature of any type of representation is questioned. And by using the means of production to emphasize the manipulations taking place in the process of visualization: by using inks with a different materiality, letting images bleed off the screen or page, and using the opacity of the paper.

I relate the position of the design of information to the notion of "situated knowledges" as advocated by American feminist scholar Donna Haraway. We need to think outside the duality of objectivity and subjectivity, Haraway claims, and instead aim "for a doctrine and practice of objectivity that privileges contestation, deconstruction, passionate construction, webbed connections, and hope for transformation of systems of knowledge and ways of seeing."[3]

Creating visualizations is a process of editing, controlling, distorting and altering data. In my work I highlight these manipulations. Noticing the misunderstandings, false expectations and presumptions embedded in the case studies, I found the research to be an endorsement of the intentional ambiguous strategies I employ in my design practice. The presentation of complex information calls for a design approach that both presents the content and questions itself, that is, questions the methods and formats employed to show it. When there is no clarity, show the opacity.

A similar ambiguous strategy can be formulated for the interpretation of visualizations. Question what a visualization looks like; doubt what looks familiar. Question what a visualization shows: what is shown, and, especially, what is not shown? Question first and foremost why it is shown and wonder if it should be shown at all.

Multiple Languages

When I started this research process, I expected to gain insights into and new ideas about the relationship between the map, its producer and user. The Blind Map and the Blue Dot became visual concepts that I developed along the way. A totally unexpected insight was the obstructing role of specialist language in advancing a field.

Specialist language employed by the practitioners of a field, like technical jargon, enables the exchange of ideas within a field, but it also functions as a form of gatekeeping for people outside the field. While the tools of graphic design have been democratized, its terminology has not. Language thus becomes an obstacle in opening up the field to new players. In this book I have tried to write texts devoid of jargon. This kind of technical language, in my opinion, is a means to exclude rather than to connect.

Technical language also complicates the re-evaluation of graphic design beyond its current conceptualization. As terminology is tied to a certain way of operating, documenting and validating, the development of alternative ways to do, see and evaluate needs to happen outside the reach of that specialist language.

Yet another way in which specialist language plays an obstructing role is how it causes miscommunication in transdisciplinary exchanges. The tradition of every professional field is embedded within the language it uses. A meeting between fields is a meeting of languages: the confrontation of different sets of concepts

3   Haraway, "Situated Knowledges: The Science Question in Feminism and the Privilege of Partial Perspective," 584–585.

173

Parallels: Social Media

Facebook launched

Instagram launched

Whatsapp launched

Facebook M

Twitter launched

Adobe DPS
Adobe Creative Cloud
iPad
iA Writer

Typekit

e-pub
Kindle
iPhone

Strava launched

TencentVide
WeChat

Kickstarter
Newspaper Club
WeTransfer
Sina Weibo
4G

GitHub
Instapaper
MagCloud

Dropbox
Tumblr

Blurb
Twitter

YouTube

Facebook
Flickr

2004   2005   2010

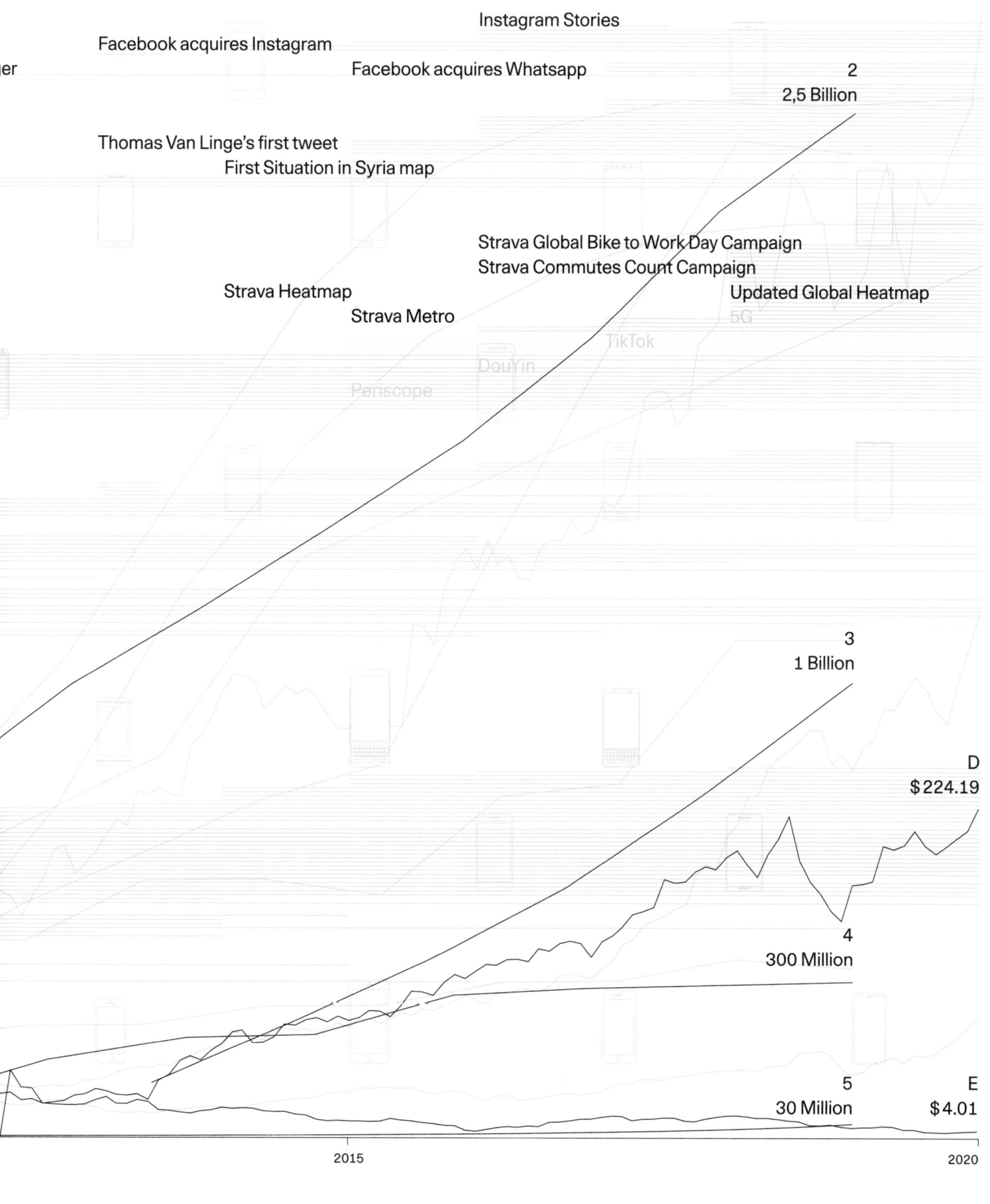

Instagram Stories

Facebook acquires Instagram

Facebook acquires Whatsapp

2

2,5 Billion

Thomas Van Linge's first tweet

First Situation in Syria map

Strava Global Bike to Work Day Campaign

Strava Commutes Count Campaign

Strava Heatmap

Updated Global Heatmap

Strava Metro

5G

DouYin

TikTok

Periscope

3

1 Billion

D

$ 224.19

4

300 Million

5

E

30 Million

$ 4.01

2015

2020

1   Monthly Google Maps users
2   Monthly Facebook users
3   Monthly Instagram users
4   Monthly Twitter users
5   Strava members
6   Worldwide sales of smartphones
7   Worldwide sales of smartphones running
    on Android

A   Adobe stock price
B   Apple stock price
C   Google stock price
D   Facebook stock price
E   Twitter stock price

Operational satellites
Satellites in reserve / Testing

and the encounter of different origins. In transdisciplinary conversations we need to acknowledge that each discipline uses a fundamentally different language.

An expanded understanding of the field, like the one I have tried to develop in this book, therefore needs to address the modes of communication used to characterize it. It has been my ambition to go one step further and also develop alternative and additional languages to characterize graphic design. In this book I have added an additional flow of information that is primarily visual. This second strand documents the same research as the texts, but uses an alternative language that requires a different mode of processing the contents: looking instead of reading.

The design of the alternative and additional visual language is informed by the "mixed methods" approach that combines different kinds of data in a single display. What appeals to me in the mixed methods approach is its aim to combine qualitative and quantitative research. It fits the subject and strategy of my investigation. On the one hand the complex data and in-depth analysis of contemporary mapmaking practices, and on the other the low-complexity data of the comparison of a large variety of graphic design tools, practices and histories. I also see a connection between mixed methods and post-representational cartography in the ability of the latter to bring together different kinds of theoretical approaches. The British geographers Rob Kitchin and Martin Dodge state that post-representational cartography offers a theoretical space where the practical applied knowledge of representational cartography and the critical thinking of more-than-representational cartography can meet.[4]

In my view, the development of alternative and complementary languages is an essential aspect of artistic research. Concerned with, and conducted by means of practice, artistic research is a type of investigation that produces two outputs, a discursive and an artistic one. In my case I have combined them into a single format, a book, but they remain two distinct flows of information that complement each other but at the same time are each an alternative incarnation of the other. I see the artistic output of artistic research projects as an argumentation embodied in an alternative language. This artistic argumentation is self-contained and complete as a report of the investigation in itself. At the same time, as an additional presentation to the discursive argumentation, this artistic line of reasoning questions the language and the format of the discursive product and all prejudices and histories contained therein. In that respect, there are similarities between artistic research and the ambiguous strategies I employ in my practice: simultaneously presenting the content and questioning it in order to highlight the fundamentally constructed nature of research.

This investigation started from a sense of urgency I experienced in my own practice. I felt the need to understand the changes that were happening in graphic design. Crucially, with the urgency came the objective to bring the insights I would obtain in the research back into my artistic practice, to use them as strategies in my design work. I was therefore as equally interested in the "how" as in the "what" of the case studies. During the research process, I realized that a purely critical position would not bring me a new awareness or a perspective that I could transport back into my practice. In thoroughly studying the technology companies I found many troubling things, but this did not result in ideas I could use in my practice. Nor did an empirical approach aimed at improving the communication help my design work, because this seemed insufficient to address the full complexity of information exchange. The insight I wanted to bring back into my practice would have to be a combination of those two approaches, questioning the hidden structures behind and showing the full scope of the data used for a visualization.

So, what insights *can* I bring back to my practice? Post-representational cartography and mixed methods appeal to me because they offer a step beyond the quantitative and qualitative research paradigms. The post-representational approach offers a space where critical and functional research methods can meet. In this way I can closely scrutinize underlying power structures, and at the same time I can study how these are supported by design choices. Similarly, but in a different way, mixed methods offer me the option to combine things, rather than to select them. For this book, it means the merging of various presentation formats. A third crucial insight is the approach of "withholding judgment" as propagated by Venturi, Scott Brown and Izenour in *Learning from Las Vegas* (1972).[5] Postponing opinion helped me see beyond the dubious surveillance tactics of technology companies to discover the deceivingly familiar formal language that was employed to camouflage commercial space as public space. Looking beyond the occasionally clumsy designs of amateur mapmakers allowed me to get to know their interesting strategies for production, distribution and exchange. All three approaches listed above are essentially methods to not choose: to combine modes of research, to combine modes of presentation and to postpone opinion. The combination of the three helps me to observe, understand and design.

Postscript: In the Future

In 1984, at the start of the process of the digitization of the means of production of visual information, the song "In the Future," by Scottish-born American songwriter, musician and artist David Byrne was performed for the first time.[6] In it there is a line that resonates with me: "In the future TV will be so good that the printed word will function as an art form only." It feels as if the line addresses me directly: I design printed matter, I have made a book about my work titled *I swear I use no art at all*, and it is me who believes that design can play a meaningful role in informing users. I read in the line from Byrne's song that technological developments will transform a field, and that this change is inevitable. Could it be possible that the fate of graphic design is to be only art? This research project has taught me to embrace ambiguity, and that the choice not to choose—by postponing judgment, by combining several research methods and by using multiple types of representation—leads to a deeper understanding of the many complexities in the production of visual information. I am convinced that a multi-layered approach is capable of coping with unforeseen changes. This is what I intended to do in this research: to open up my field, to make it more inclusive in terms of the types of works produced, the variety of practices and the kinds of languages used to characterize it. In other words, in this book I have attempted to open up the thinking about graphic design so that it can become future-proof.

4   Kitchin and Dodge, "Rethinking Maps," 337.
5   Venturi, Scott Brown and Izenour, *Learning from Las Vegas: The Forgotten Symbolism of Architectural Form.*
6   Byrne's song "In the Future" was part of *The Knee Plays,* a section of the larger opera *the CIVIL wars: a tree is best measured when it is down,* by American playwright and theater artist Robert Wilson. *The Knee Plays* premiered in April 1984, three months after the Apple Macintosh computer went on sale. Incidentally, the premiere of *The Knee Plays* took place at the Walker Art Center, Minneapolis, where the aforementioned Andrew Blauvelt was design director and curator of, among others, the *Graphic Design: Now in Production* (2011) exhibition.

A
F
P
Wacom
Linux
Digital stock ph

Adobe Illustrator
Adobe Photoshop
QuarkXPress
Screen reader

Adobe founded in Mountain View, California, US
Adobe PageMaker
LaserWriter desktop printer
Macintosh computer
Apple founded in Cupertino, California, US InkJet inkjet printer
Apple goes public
News reader (usenet)
Copyleft   Apple licenses PostScript
Digital camera   PostScript first appears
Microsoft   Korean Air Lines Flight 007 shot down
                       Soviet Union launches GLONASS
Desktop publishing   The Gulf War, firs
Computer generated image (CGI) programs   US launches GPS. Restricted to military use only
Digital fonts
Geographic information systems (GIS)
Mouse

| 1970 | 1975 | 1980 | 1985 | 1990 |

Digitization   Selective availability   D

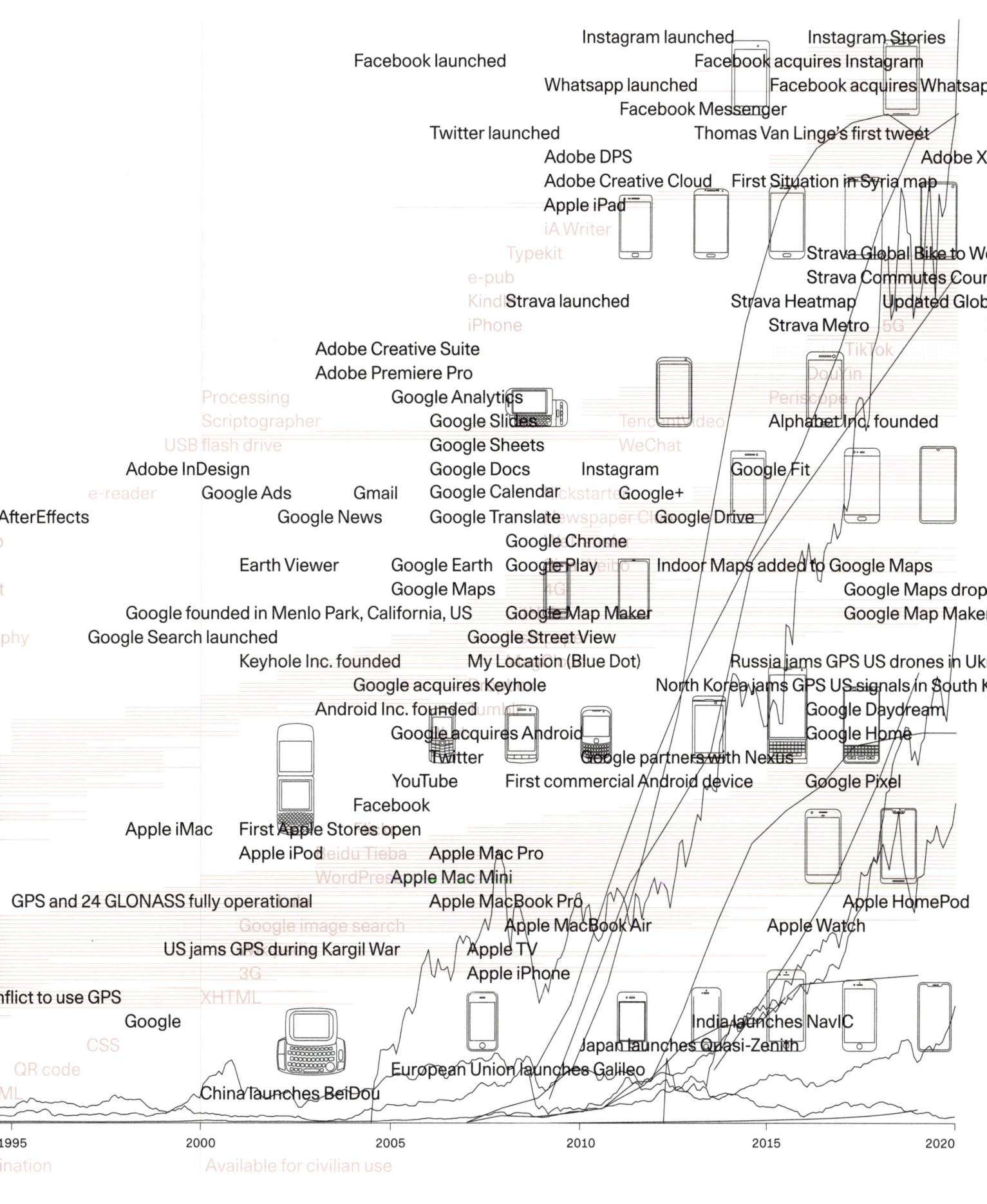

1	Monthly Google Maps users	A	Adobe stock price
2	Monthly Facebook users	B	Apple stock price
3	Monthly Instagram users	C	Google stock price
4	Monthly Twitter users	D	Facebook stock price
5	Strava members	E	Twitter stock price
6	Worldwide sales of smartphones		
7	Worldwide sales of smartphones running on Android		

Operational satellites
Satellites in reserve / Testing

Bibliography

Barthes, Roland, "The Death of the Author" (1967), in *Image Music Text* (London: Fontana Press, 1977), 142–148.

Bayer, Herbert, ed., *World Geo-Graphic Atlas: A Composite of Man's Environment* (Chicago: Container Corporation of America, 1953).

Belz, Corrina, *Gerhard Richter Painting,* documentary, 2011.

Benjamin, Walter, "The Author as Producer" (1934), in *The Work of Art in the Age of Its Technological Reproducibility, and Other Writings on Media*, eds. Michael W. Jennings, Brigid Doherty and Thomas Y. Levin (Cambridge, MA: Harvard University Press, 2008), 79–95.

Bertin, Jacques, *Semiology of Graphics: Diagrams Networks Maps* (1967; Redlands: Esri Press, 2011).

Black, Alison, Paul Luna, Ole Lund, and Sue Walker, eds., *Information Design: Research and Practice* (Oxon: Routledge, 2017).

Blauvelt, Andrew, "Towards Critical Autonomy, or Can Graphic Design Save Itself?" (2003), in *Looking Closer 5: Critical Writings on Graphic Design*, eds. Michael Bierut, William Drenttel and Stephen Heller (New York: Allworth Press, 2006), 8–11.

Blauvelt, Andrew, "Tool (Or, Post-production for the Graphic Designer)," in *Graphic Design: Now in Production*, eds. Andrew Blauvelt and Ellen Lupton (Minneapolis: Walker Art Center, 2011), 23–31.

Blauvelt, Andrew and Ellen Lupton, eds., *Graphic Design: Now in Production* (Minneapolis: Walker Art Center, 2011).

Blauvelt, Andrew, "Graphic Design: Discipline, Medium, Practice, Tool, or Other?," lecture filmed 11 May 2013 at "counter/point: The 2013 D-Crit Conference," School of Visual Arts, New York, https://vimeo.com/66385792.

Boekraad, Hugues, "Graphic Design as Visual Rhetoric. Principles for Design Education," in *Copy Proof: A New Method for Design Education*, eds. Edith Gruson and Gert Staal (Rotterdam: 010 Publishers, 2000), 4–14.

De Bondt, Sara, and Catherine De Smet, eds., *Graphic Design: History in the Writing (1983–2011)* (London: Occasional Papers, 2012).

Brotton, Jerry, *A History of the World in Twelve Maps* (London: Penguin Books, 2012).

Buchler, Justus, ed., *Philosophical Writings of Peirce* (New York: Dover Productions, 1955).

Byrne, David, "In the Future," in *The Knee Plays* (1984; New York: Nonesuch Records, 1985).

Caquard, Sébastien, "A Post-Representational Perspective on Cognitive Cartography," *Progress in Human Geography* 39, no. 2 (2015), 225–235.

Crampton, Jeremy W., and John Krygier, "An Introduction to Critical Cartography," *ACME: An International E-Journal for Critical Geographies* 4, no. 1 (2005), 11–33.

Creswell, John W., and Vicki L. Plano Clark, *Designing and Conducting Mixed Methods Research*, 2nd edition (Los Angeles: SAGE, 2011).

Del Casino Jr., Vincent J., and Stephen P. Hanna, "Beyond the 'Binaries': A Methodological Intervention for Interrogating Maps as Representational Practices," *ACME: An International E-Journal for Critical Geographies* 4, no. 1 (2005), 34–56.

Demoule, Jean-Paul, Dominique Garcia and Alain Schnapp, eds., *Une histoire des civilisations: Comment l'archéologie bouleverse nos connaissances* (Paris: Éditions La Découverte/Inrap, 2018).

"Design graphique et recherches en sciences sociales: Jacques Bertin et le Laboratoire de Graphique. EHESS 1954-2000," conference, École des Hautes Études en Sciences Sociales (EHESS), Paris, 13 November 2017. Conference website, accessed 01 March 2018, http://retrospective-bertin.ehess.fr.

Dewey, John, *The Quest for Certainty: A Study of the Relation of Knowledge and Action* (London: George Allen and Unwin Limited, 1930).

Drucker, Johanna, and Emily McVarish, eds., *Graphic Design History: A Critical Guide*, 2nd edition (Boston: Pearson, 2013)

Drucker, Johanna, *Graphesis: Visual Forms of Knowledge Production* (Cambridge, MA: Harvard University Press, 2014).

Dunne, Anthony, *Hertzian Tales: Electronic Products, Aesthetic Experience, and Critical Design* (Cambridge, MA: The MIT Press, 1999).

Eckert, Denis, and Laurent Jégou, "Quel planisphère de références pour Google Maps," *Mappe-monde* 92, no. 4 (2008), accessed 3 May 2019, http://mappemonde.mgm.fr/num20/internet/int08401.html.

Eckert, Denis, "Is Innovation in Cartography a Mere Illusion?," lecture at "Mappings as Joint Spatial Display" conference, TU Berlin, 30 November 2018.

Eskilson, Stephen J., *Graphic Design: A History*, 2nd edition (London: Laurence King, 2012).

Foucault, Michel, "What Is an Author?" (1969) *Screen* 20, no. 1 (Spring 1979), 13–34.

Friendly, Michael, "A Brief History of Data Visualization," in *Handbook of Data Visualization*, eds. Chun-houh Chen, Wolfgang Härdle, Antony Unwin (Berlin: Springer-Verlag, 2008), accessed 31 August 2019, http://datavis.ca/papers/hbook.pdf.

Goggin, James, "Practice from Everyday Life: Defining Graphic Design's Expansive Scope by its Quotidian Activities" (2009), in *Graphic Design: Now in Production*, eds. Andrew Blauvelt and Ellen Lupton (Minneapolis: Walker Art Center, 2011), 55–56.

Hacking, Ian, *Representing and Intervening: Introductory Topics in the Philosophy of Natural Science* (New York: Cambridge University Press, 1983).

Haraway, Donna, "Situated Knowledges: The Science Question in Feminism and the Privilege of Partial Perspective," *Feminist Studies* 14, no. 3 (1988), 575–599.

Harley, J. B., "Deconstructing the Map" (1989), in *The Map Reader: Theories of Mapping Practice and Cartographic Representation*, eds. Martin Dodge, Rob Kitchin and Chris Perkins (Chichester: John Wiley & Sons, 2011), 56–64.

Hollis, Richard, "Have You Ever Really Looked at This Poster?" (1994), in *Graphic Design: History in the Writing (1983–2011)*, eds. Sara De Bondt and Catherine De Smet (London: Occasional Papers, 2012), 73–75.

Hollis, Richard, *Graphic Design: A Concise History* (London: Thames & Hudson, 2001).

Huffschmid, Anne, "Reconstructing Conflict: Mapping as Materialization of Contested Memories and Invisibilized Crime," lecture at "Mappings as Joint Spatial Display" conference, TU Berlin, 30 November 2018.

Jubert, Roxane, *Typography and Graphic Design: From Antiquity to the Present* (2005; Paris: Flammarion, 2006).

Kitchin, Rob, and Martin Dodge, "Rethinking Maps," *Progress in Human Geography* 31, no. 3 (2007), 331–344.

Kitchin, Rob, "The Transformation of Cartographic Thought," lecture filmed September 2012 at "Mapping Humans: From Body to Cosmos: An International Conference," Oxford, accessed 9 November 2018, http://www.pulse-project.org/node/540.

Kurgan, Laura, *Close Up at a Distance: Mapping, Technology, and Politics* (New York: Zone Books, 2013).

Lommen, Mathieu, *Het boek van het gedrukte boek: Een visuele geschiedenis* (Amsterdam: Amsterdam University Press, 2012).

Lupton, Deborah, "Self-Tracking Modes: Reflexive Self-Monitoring and Data Practices," paper for the "Imminent Citizenships: Personhood and Identity Politics in the Informatic Age" workshop, 27 August 2014, ANU, Canberra, accessed 25 November 2019, https://www.academia.edu/attachments/34502881/download_file?st=MTU3NDY5NzQ3NCw2Mi4yNTEuMTcuMTYzLDQ4MTg2MTM3&s=profile&ct=MTU3NDY5NzQ0OCwxNTc0Njk3NDc5LDQ4MTg2MTM3.

Lupton, Ellen, "The Designer as Producer" (1998), in *Graphic Design: Now in Production*, eds. Andrew Blauvelt and Ellen Lupton (Minneapolis: Walker Art Center, 2011), 13.

Lyon, David, "Liquid Surveillance: The Contribution of Zygmunt Bauman to Surveillance Studies," *International Political Sociology* 4 (2010), 325–338.

Manovich, Lev, *Software Takes Command* (2013; New York: Bloomsbury Academic, 2014).

"Mappings as Joint Spatial Display," conference, Haus der Kulturen der Welt, 29 November 2018, and Technische Universität Berlin, 30 November 2018. Conference websites, accessed 2 March 2019, https://www.hkw.de/en/programm/projekte/veranstaltung/p_144947.php and http://www.cud.tu-berlin.de/topics/spatial-commons-8.

Meggs, Philip B., and Alston W. Purvis, *Meggs' History of Graphic Design*, 5th edition (Hoboken: John Wiley & Sons, 2012).

Neumann, Joachim, ed., *Enzyklopädisches Wörterbuch Kartographie in 25 Sprachen* (Munich: K.G. Saur Verlag GmbH & Co., 1997).

Offenhuber, Dietmar, "Maps of Daesh: The Cartographic Warfare Surrounding Insurgent Statehood," *GeoHumanities* 4 (2018), accessed 31 May 2019, https://www.academia.edu/35536337/Maps_of_Daesh_The_Cartographic_Warfare_Surrounding_Insurgent_Statehood.

Parks, Lisa, "Plotting the Personal: Global Positioning Satellites and Interactive Media," *Cultural Geographies* 8, no. 2 (2001), 209–222.

Perec, Georges, *An Attempt at Exhausting a Place in Paris* (1975; Cambridge: Wakefield Press, 2010).

Pickles, John, *A History of Spaces: Cartographic Reason, Mapping and the Geo-Coded World* (London: Routledge, 2004).

Poynor, Rick, *No More Rules: Graphic Design and Postmodernism* (London: Laurence King Publishing, 2003).

Richter, Gerhard, and Hans Ulrich Obrist, *The Daily Practice of Painting: Writings and Interviews, 1962-1993* (Cambridge, MA: MIT Press, 1995).

Rock, Michael, "The Designer as Author," *Eye* 20 (Spring 1996), accessed 25 September 2016, http://www.eyemagazine.com/feature/article/the-designer-as-author.

Rock, Michael, "Fuck Content" (2005), in *Graphic Design: Now in Production*, eds. Andrew Blauvelt and Ellen Lupton (Minneapolis: Walker Art Center, 2011), 15.

San-Antonio-Gómez, Carlos, C. Velilla and Francisco Manzano-Agugliaro, "Tomás López's Geographic

Barthes, "The Death of the Author"   Rock, "The Des

Foucault, "What is an Author?"   L

Hollis, "Have You Ever R

Bertin, *Semiology of Graphics: Diagrams Networks Maps*

Harley, "Deconstructing the Map"

Venturi, Scott Brown & Izenour, *Learning from Las Vegas*

Byrne, "In the Future"

Hacking, *Representing and Intervening*

Perec, *An Attempt at Exhausting a Place in Paris*   Haraway, "Situated Knowledges"   D

Zuboff, *In the Age of the Smart Machine*

| 1970 | 1975 | 1980 | 1985 | 1990 | 1995 |

In this chronological overview, the different publications in the bibliography are
divided horizontally by publication date and vertically by field of knowledge.

as Author"

"The Designer as Producer"   De Smet, "*Pussy Galore* and Buddha of the Future"

Goggin, "Practice from Everyday Life"

Blauvelt, "Towards Critical Autonomy, or Can Graphic Design Save Itself?"

Poynor, *No More Rules*

Blauvelt & Lupton, *Graphic Design: Now in Production*

Blauvelt, "Tool (Or, Post-production for the Graphic Designer)"

ekraad, "Graphic Design as Visual Rhetoric"   Blauvelt, "Graphic Design: Discipline, Medium, Practice, Tool, or Other?"

ooked at This Poster?"

Rock, "Fuck Content"   De Bondt & De Smet, *Graphic Design: History in the Writing (1983–2011)*

Eskilson, *Graphic Design: A History*

Meggs & Purvis, *Meggs' History of Graphic Design*

Drucker & McVarish, *Graphic Design History: A Critical Guide*

Hollis, *Graphic Design: A Concise History*

Jubert, *Typography and Graphic Design*

"Jacques Bertin et le Laboratoire de Graphique"

Drucker, *Graphesis: Visual Forms of Knowledge Production*

Friendly, "A Brief History of Data Visualization"

Wilkinson & Friendly, "The History of the Cluster Heat Map"

Tufte, *The Cognitive Style of PowerPoint*   "Visualizing Knowledge 2019"

Pickles, *A History of Spaces*   Eckert, "Is Innovation in Cartography a Mere Illusion?"

Kitchin, "The Transformation of Cartographic Thought"

Crampton & Krygier, "An Introduction to Critical Cartography"

Offenhuber, "Maps of Daesh"

Kitchin & Dodge, "Rethinking Maps"

Del Casino Jr. & Hanna, "Beyond the 'Binaries'"

Brotton, *A History of the World in Twelve Maps*

Caquard, "A Post-Representational Perspective on Cognitive Cartog

Kurgan, *Close Up at a Distance*

Weizman, *Forensic Architecture*

Lupton, "Self-Tracking Modes"

Lyon, "Liquid Surveillance"

Huffschmid, "Reconstructing Conflict"

"Mappings as Joint Spatial Display"

Creswell & Plano Clark, *Designing and Conducting Mixed Methods Research*

Parks, "Plotting the Personal"   Zuboff, "Google as a Fortune Teller"

Hertzian Tales   Manovich, *Software Takes Command*

2005   2010   2015   2020

Atlas of Spain in the Peninsulan War: A Methodology for Determining Errors," *Survey Review* 43, no. 319 (2011), 30–44.

Simpson, Deane, Kathrin Gimmel, Anders Lonka, Marc Jay and Joost Grootens, eds., *Atlas of the Copenhagens* (Berlin: Ruby Press, 2018).

De Smet, Catherine, "*Pussy Galore* and Buddha of the Future: Women, Graphic Design, etc." (2009), in *Graphic Design: History in the Writing (1983–2011)*, eds. Sara De Bondt and Catherine De Smet (London: Occasional Papers, 2012), 251–255.

Tufte, Edward R., *The Cognitive Style of PowerPoint* (Cheshire: Graphics Press, 2006).

Venturi, Robert, Denise Scott Brown, and Steven Izenour, *Learning from Las Vegas: The Forgotten Symbolism of Architectural Form* (1972; Cambridge, MA: MIT Press, 1977).

"Visualizing Knowledge 2019" conference, Aalto University, Helsinki, 10 May 2019. Conference website, accessed 25 August 2019, https://vizknowledge.aalto.fi.

Weizman, Eyal, *Forensic Architecture: Violence at the Threshold of Detectability* (Brooklyn: Zone Books, 2017).

Wilkinson, Leland, and Michael Friendly, "The History of the Cluster Heat Map," *The American Statistician* 63 (2009), 179–184.

Zuboff, Shoshana, *In the Age of the Smart Machine: The Future of Work and Power* (New York: Basic Books, 1988).

Zuboff, Shoshana, "The Surveillance Paradigm: Be the Friction—Our Response to the New Lords of the Ring," *Frankfurter Algemeine Zeitung*, 25 June 2013, accessed 2 April 2018, http://www.faz.net/aktuell/feuilleton/the-surveillance-paradigm-be-the-friction-our-response-to-the-new-lords-of-the-ring-12241996.html?printPagedArticle=true#pageIndex_0.

Zuboff, Shoshana, "Google as a Fortune Teller: The Secrets of Surveillance Capitalism," *Frankfurter Allgemeine Zeitung*, 5 March 2016, accessed 2 April 2018, http://www.faz.net/aktuell/feuilleton/debatten/the-digital-debate/shoshana-zuboff-secrets-of-surveillance-capitalism-14103616.html?printPagedArticle=true.

Additional sources
Blue Dot case study

"A fresh new look for the Maps API, for all one million sites," Google Geo Developers Blog, blog post, 15 May 2013, accessed 3 May 2019, https://web.archive.org/web/20131128131730/http://googlegeodevelopers.blogspot.com/2013/05/a-fresh-new-look-for-maps-api-for-all.html.

Ball, James, "Angry Birds and 'Leaky' Phone Apps Targeted by NSA and GCHQ for User Data," *The Guardian*, 28 January 2014, accessed 29 November 2019, https://www.theguardian.com/world/2014/jan/27/nsa-gchq-smartphone-app-angry-birds-personal-data.

Buczkowski, Aleks, "Google Maps Get Redesign of the Blue Dot Showing Your Position," *Geo Awesomeness*, 23 September 2016, accessed 17 May 2019, https://geoawesomeness.com/google-maps-get-redesign-of-the-blue-dot-showing-your-position.

Chu, Mike, "New Magical Blue Circle on Your Map," blog post, 28 November 2007, accessed 01 October 2015, http://googlemobile.blogspot.nl/2007/11/new-magical-blue-circle-on-your-map.html.

"Fade to White," *The Observatory* podcast episode 104, 17 May 2019, accessed 17 May 2019, https://designobserver.com/article.php?id=40070.

"Google Maps for Mobile with My Location (beta)," Google, animation, 27 November 2007, accessed 16 May 2019, https://www.youtube.com/watch?v=v6gqipmbcok.

"LinkedIn profile Jonathan Lee," accessed 17 May 2019, https://www.linkedin.com/in/hifromjonathan/.

"LinkedIn profile Sanjay Mavinkurve," accessed 3 March 2019, https://www.linkedin.com/in/sanjay-mavinkurve-3b819b1/.

Maney, Kevin, "Tiny Tech Company Awes Viewers," *USA Today*, 21 March 2003, accessed 12 May 2019, https://usatoday30.usatoday.com/tech/news/techinnovations/2003-03-20-earthviewer_x.htm.

O'Beirne, Justin, "What Happened to Google Maps?," blog post (2016), accessed 12 May 2019, https://www.justinobeirne.com/what-happened-to-google-maps.

Orerskovic, Alexei, "Sundar Pichai Just Hinted at How Google Will Make Money from Maps, and it Sounds Like Lots of Ads," in *Business Insider*, 28 April 2017, accessed 4 May 2019, https://www.businessinsider.nl/sundar-pichai-hints-at-ads-in-google-maps-2017-4/?international=true&r=US.

Panko, Riley, "The Popularity of Google Maps: Trends in Navigation Apps in 2018," *The Manifest*, 10 July 2018, accessed 3 May 2019, https://themanifest.com/app-development/popularity-google-maps-trends-navigation-apps-2018.

Pichai, Sundar, "Google Keynote," lecture at "I/O 2017" conference, Shoreline Amphitheatre, Mountain View, 17–19 May 2017, accessed 3 May 2019, https://www.youtube.com/watch?v=vWLcyFtni6U.

Siegler, M. C., "Google Maps for Mobile Crosses 200 Million Installs; In June It Will Surpass Desktop Usage," *TechCrunch*, 25 May 2011, accessed 12 May 2019, https://techcrunch.com/2011/05/25/google-maps-for-mobile-stats/?guccounter=1.

Usborne, Simon, "Disputed Territories: Where Google Maps Draws the Line," *The Guardian*, 10 August 2016, accessed 12 May 2019, https://www.theguardian.com/technology/shortcuts/2016/aug/10/google-maps-disputed-territories-palestineishere.

Additional sources
Strava Global Heatmap case study

Bradshaw, Tim, "Under Armour Snaps up Fitness Apps," *Financial Times*, 5 February 2015, accessed 6 April 2018, https://www.ft.com/content/2eed0aac-acc7-11e4-beeb-00144feab7de.

Clement, J. (2019), "Number of Monthly Active Facebook Users Worldwide as of 2nd Quarter 2019," *Statista* 9 August 2019, accessed 24 August 2019, https://www.statista.com/statistics/264810/number-of-monthly-active-facebook-users-worldwide.

"#CommutesCount," Strava Metro, animation, 24 April 2016, accessed 24 August 2019, https://www.youtube.com/watch?v=LYUmhIAwuRA.

"Commutes Count," Strava, animation, 3 May 2017, accessed 24 August 2019, https://www.youtube.com/watch?v=WcGoR5RWQNU.

"Digital Dystopia: Tech Slavery and the Death of Privacy," *Chips with Everything* podcast, 12 January 2018, accessed 5 March 2018, https://www.theguardian.com/technology/audio/2018/jan/12/digital-dystopia-end-of-privacy-tech-podcast.

Dissinger, Kaleb, "GPS Goes to War: The Global Positioning System in Operation Desert Storm," US Army, accessed 31 March 2018, https://www.army.mil/article/7457/gps_goes_to_war_the_global_positioning_system_in_operation_desert_storm.

Hern, Alex, "Fitness Tracking App Gives away Location of Secret US Army Bases," *The Guardian*, 28 January 2018, accessed 24 August 2019, https://www.theguardian.com/world/2018/jan/28/fitness-tracking-app-gives-away-location-of-secret-us-army-bases.

Lassiter III, Joseph B., William A. Sahlman, and Sid Misra, "Strava. Harvard Business School Case 814-055," February 2014 (Revised August 2016).

Lilley, Kevin, "20,000 Soldiers Tapped for Army Fitness Program's 2nd Trial," *Army Times*, 27 July 2015, accessed 5 April 2018, https://www.armytimes.com/news/your-army/2015/07/27/20000-soldiers-tapped-for-army-fitness-program-s-2nd-trial.

Malouff, Dan, "Heat maps show where people bike... or at least, where affluent people exercise by bike," *Greater Greater Washington*, 12 May 2014, accessed 6 April 2018, https://ggwash.org/view/34716/heat-maps-show-where-people-bike-or-at-least-where-affluent-people-exercise-by-bike.

Olson, Parmy, "Why Google's Waze Is Trading User Data with Local Governments," *Forbes*, 7 July 2014, accessed 6 April 2018, https://www.forbes.com/sites/parmyolson/2014/07/07/why-google-waze-helps-local-governments-track-its-users/#494ce86f39ba.

Robb, Drew, "The Global Heatmap, Now 6× Hotter," Strava, 1 November 2017, accessed 3 March 2018, https://medium.com/strava-engineering/the-global-heatmap-now-6x-hotter-23fc01d301de.

Ruser, Nathan (@Nrg8000), "Strava released their global heatmap: 13 trillion GPS points from their users (turning off data sharing is an option). It looks very pretty, but not amazing for Op-Sec [JG: operation security]. US Bases are clearly identifiable and mappable," Twitter message, 27 January 2018, accessed 3 March 2018, https://twitter.com/Nrg8000/status/957318498102865920.

Shaw, Mark, "The Story behind Strava Metro," lecture at "FutureStack14" conference, Fort Mason Center, San Francisco, 8–9 October 2014, accessed 6 April 2018, https://www.youtube.com/watch?v=0zAG3j5JAAc.

Strava, "About Us," accessed 6 April 2018, https://www.strava.com/about.

"Strava Global Heatmap," Strava, accessed 31 March 2018, https://www.strava.com/heatmap.

"Strava Metro," website, accessed 6 April 2018, https://metro.strava.com.

Tenetz, Antti, "Tracing—Jalestaa," website, accessed 25 August 2019, http://www.tenetz.com.

Tilghman, Andrew, "The US Military Has a Huge Problem with Obesity and It's Only Getting Worse,"

Military Times, 11 September 2016, accessed 5 April 2018, https://www.militarytimes.com/news/your-military/2016/09/11/the-u-s-military-has-a-huge-problem-with-obesity-and-it-s-only-getting-worse.

"Under Armour Reports Full Year Net Revenues Growth Of 32%; Announces Creation of World's Largest Digital Health and Fitness Community," *PR Newswire*, 4 February 2015, accessed 6 April 2018, https://www.prnewswire.com/news-releases/under-armour-reports-full-year-net-revenues-growth-of-32-announces-creation-of-worlds-largest-digital-health-and-fitness-community-300030909.html.

United States Department of Commerce, National Oceanic and Atmospheric Administration, "GPS & Selective Availability Q&A" (date unknown), accessed 31 March 2018, https://web.archive.org/web/20050921115614/http://ngs.woc.noaa.gov/FGCS/info/sans_SA/docs/GPS_SA_Event_QAs.pdf.

United States Office of Science and Technology Policy, National Security Council, "Press Release US Global Positioning System Policy," 29 March 1996, accessed 31 March 2018, https://clintonwhitehouse4.archives.gov/textonly/WH/EOP/OSTP/html/gps-factsheet.html.

United States Patent and Trademark Office, "Heatmaps," 3 March 1998, accessed 6 April 2018, http://tsdr.uspto.gov/documentviewer?caseId=sn75263259&docId=UNC20051018111828#docIndex=0&page=1.

"What is GPS," Garmin, accessed 2 April 2018, https://www8.garmin.com/aboutGPS/.

Additional sources
Situation in Syria case study

"18-jarige kaartenmaker verovert de wereld," NOS op 3, 16 June 2015, accessed 11 August 2019, https://nos.nl/op3/artikel/2041770-18-jarige-kaartenmaker-verovert-de-wereld.html.

BAK, "Forensic Justice," accessed 8 June 2019, https://www.bakonline.org/program-item/forensic-justice.

Barochová, Anna, "Válku v Sýrii mapují amatérští kartografové, ofenzivy ukazují online," iDNES.cz, 6 April 2016, accessed 15 March 2020, https://www.idnes.cz/zpravy/zahranicni/syrie-mapy-kartograf-konflikt-mapovani.A160414_171101_zahranicni_aba.

"Botsingen tussen politie en betogers op bezet vliegveld Hongkong," RTL Nieuws, 14 August 2019, accessed 17 August 2019, https://www.rtlnieuws.nl/nieuws/buitenland/artikel/4812801/oproerpolitie-bestormt-bezet-vliegveld-hongkong.

Cérez, Gael, and Chris O'Brien, "Watching War: Online Mapmakers Chart Syrian Conflict," National Geographic, 8 April 2016, accessed 18 December 2019, https://www.nationalgeographic.com/news/2016/04/160408-online-mapmakers-chart-syrian-conflict/.

Delille, Benjamin, "Le meilleur cartographe de la situation en Syrie est un lycéen hollandais qui ne s'y est jamais rendu," Slate FR, 26 August 2015, accessed 18 December 2019, http://www.slate.fr/story/106051/meilleur-cartographe-syrie-lyceen-hollandais.

"Deze Nederlander van 20 volgt Islamitische Staat op de voet," Nos op 3, 9 July 2015, accessed 11 August 2019, https://www.youtube.com/watch?v=E08-JX2NrpA.

Egberts, Thom, "18-jarige Amsterdammer brengt feilloos Syrische strijdgroepen in kaart," *Het Parool*, 20 June 2015, accessed 11 August 2019, https://www.parool.nl/nieuws/18-jarige-amsterdammer-brengt-feilloos-syrische-strijdgroepen-in-kaart~bf38ca84.

"Estudiante de secundaria mapea al Estado Islámico," Newsweek México, 19 April 2018, accessed 18 December 2019, https://newsweekespanol.com/2015/06/estudiante-de-secundaria-mapea-al-estado-islamico/.

Février, Renaud, "A 19 ans, il est le meilleur cartographe du conflit syrien," L'Obs, 11 September 2015, accessed 11 August 2019, https://www.youtube.com/watch?v=KR_bxt7d5YQ.

Forensic Architecture, "About," accessed 8 June 2019, https://forensic-architecture.org/about/agency.

Forensic Architecture, "Airstrikes on M2 Hospital" accessed 7 September 2019, https://forensic-architecture.org/investigation/airstrikes-on-m2-hospital.

Higgins, Eliot, "Identifying Government Positions during the August 21st Sarin Attacks," Bellingcat website, 15 July 2014, accessed 1 September 2019, https://www.bellingcat.com/news/mena/2014/07/15/identifying-government-positions-during-the-august-21st-sarin-attacks.

"Hoorzitting/rondetafelgesprek Nederlandse steun aan gewapende Syrische oppositie," Tweede Kamer der Staten Generaal, 27 September 2018, accessed 11 August 2018, https://www.tweedekamer.nl/debat_en_vergadering/commissievergaderingen/details?id=2018A03897.

Huchon, Thomas, "Carte jeune pour la Syrie," Spicee, 5 September 2015, accessed 18 December 2019, https://www.spicee.com/fr/program-guest/carte-jeune-pour-la-syrie-532.

van Huët, Bob, "Thomas (19) zit IS op de hielen voor de BBC," AD, 1 September 2015, accessed 18 December 2019, https://www.ad.nl/buitenland/thomas-19-zit-is-op-de-hielen-voor-de-bbc~affe59fc/?referrer=https://www.google.com/.

Huffschmid, Anne, and Marianne Braig, "'Knochenlesen' als Grenzüberschreitung. Forensische Anthropologie als Beitrag zur Gewaltverarbeitung und transnationaler Wissenstransfer, am Beispiel des argentinischen EAAF (Mexiko, Spanien)," web page (Freie Universität Berlin, 2015), accessed 8 September 2019, https://www.lai.fu-berlin.de/disziplinen/politikwissenschaft/forschung/forschungsprojekte/knochenlesen-als-grenzueberschreitung/index.html.

Keultjes, Hanneke, "Schone handen houden was onmogelijk in Syrië," AD, 28 September 2018, accessed 8 September 2019, "https://www.ad.nl/politiek/schone-handen-houden-was-onmogelijk-in-syrie~aeb4f6bf.

Koens, Olaf, "Deze 19-jarige Amsterdammer maakt kaarten van conflict Syrië," RTL Nieuws, 23 November 2015, accessed 11 August 2019, https://www.youtube.com/watch?v=oE9GAl3_P7c.

Kuntz, Katrin, "The Dutch Teen Who Maps the Jihadists," Spiegel Online, 19 August 2015, accessed 11 August 2019, https://www.spiegel.de/international/world/how-thomas-van-linge-mapped-islamic-state-a-1048665.html.

"Le meilleur cartographe de Daesh au monde... serait un ado hollandais," RT France, 21 August 2015, accessed 18 December 2019, https://francais.rt.com/international/5887-meilleur-cartographe-daesh-monde-est-adolescent-hollandais.

Ley, Julia, "Kartograf des Krieges," *Süddeutsche Zeitung*, 8 September 2015, accessed 18 December 2019, https://www.sueddeutsche.de/politik/buergerkrieg-in-syrien-kartograf-des-krieges-1.2634480.

"Les cartes du conflit syrien d'un adolescent deviennent virales," The Times of Israël, 9 September 2015, accessed 18 December 2019, https://fr.timesofisrael.com/les-cartes-du-conflit-syrien-dun-adolecent-deviennent-virales/.

"Le dessous des cartes en Syrie, c'est lui," L'Obs, 14 September 2015, accessed 18 December 2019, https://www.nouvelobs.com/video/20150911.OBS5686/video-le-dessous-des-cartes-en-syrie-c-est-lui.html.

van Linge, Thomas (@ThomasVLinge), "IMPORTANT: map about the current situation in #syria. green = regime, brown = #FSA, blue = contested," Twitter message, 24 June 2013, accessed 16 June 2019, https://twitter.com/ThomasVLinge/status/349243455425368064.

"Nederlandse kaartenmaker (19) toont Poetin's ongelijk," Oneworld, 27 October 2015, accessed 18 December 2019, https://www.oneworld.nl/achtergrond/nederlandse-kaartenmaker-19-toont-poetins-ongelijk/.

"Ontdek de Creative Cloud-ervaring," Adobe, accessed 16 August 2019, https://www.adobe.com/nl/creativecloud/plans.html?promoid=P3KMQYMW&mv=other.

Ray, Michael, "8 Deadliest Wars of the 21st Century," *Encyclopaedia Britannica*, accessed 16 June 2019, https://www.britannica.com/list/8-deadliest-wars-of-the-21st-century.

Ricciardelli, Alvaro A., "This Teenager Maps the Syrian War from His Bedroom," BBC Trending, 1 September 2015, accessed 11 August 2019, https://www.bbc.com/news/blogs-trending-34116366.

Schatz, Bryan, "These Digital Sleuths Are Sticking It to ISIS and the Kremlin," *Mother Jones*, 23 February 2016, accessed 18 December 2019, https://www.motherjones.com/politics/2016/02/bellingcat-intelligence-analysts-ukraine-russia-isis/.

"Thomas van Linge," Canvas, 24 September 2015, accessed 11 August 2019, https://www.canvas.be/video/terzake/najaar-2015/donderdag-24-september-2015/thomas-van-linge.

"Thomas van Linge (20) brengt conflict in het Midden-Oosten in kaart," Oemma Nieuws, 17 September 2016, accessed 11 August 2019, https://www.youtube.com/watch?v=1Gg2NGT3wRE&t=463s.

De Weerd, Remmelt, and Henk van Houtum, "Waarom je de macht van Islamitische Staat nooit in één kaart kunt vatten," *De Correspondent,* 10 July 2015, accessed 11 August 2019, https://decorrespondent.nl/3067/waarom-je-de-macht-van-is-lamitische-staat-nooit-in-een-kaart-kunt-vatten/339025036009-a7fd4731.

Westcott, Lucy, "The High School Student Who Maps ISIS's Lightning-Quick Advance," *Newsweek*, 13 June 2015, accessed 11 August 2019, https://www.newsweek.com/dutch-high-school-student-maps-isiss-terrifying-advance-syria-and-iraq-342604.

Index

Biography

Joost Grootens is a graphic designer, educator and researcher. His studio SJG designs books, maps, typefaces, spatial installations and digital information environments for publishers, including Lars Müller Publishers, nai010 publishers, Park Books, Phaidon press, Van Dale Publishers; for educational and research institutes such as ETH Zürich, Future Cities Laboratory Singapore, KADK Copenhagen, Delft University of Technology; and for museums, among which Museum Boijmans Van Beuningen Rotterdam, Stedelijk Museum Amsterdam, Van Abbemuseum Eindhoven.

SJG won numerous prizes for their designs. Among them the "Goldene Letter"and two Gold Medals in the Best Book Design from all over the World competition in Leipzig, twice the Dutch Design Award for Graphic Design, and the Rotterdam Design Prize 2009. A monograph about SJG's work titled *I swear I use no art at all* was published by 010 Publishers in 2010.

Grootens is head of the master programme Information Design at Design Academy Eindhoven and is a university lecturer and researcher at Leiden University's Academy of Creative and Performing Arts. He also holds teaching positions at the Urbanism and Societal Change master of the Royal Danish Academy of Fine Arts, School of Architecture, Copenhagen, and the Editorial Design master of the Istituto Superiore per le Industrie Artistiche, Urbino.

Grootens studied architectural design at Gerrit Rietveld Academy Amsterdam and obtained his doctorate at Leiden University. His research addresses the transformation of the fields and practices of graphic design and mapmaking resulting from technological changes in tools to record, create, edit, produce and disseminate visual information.

Acknowledgments

As this book is rooted in, and connected to, my different practices in design, education and research, there are many people to thank.

First, I would like to express my gratitude to my supervisor Janneke Wesseling, whose incisive guidance, great involvement and encouragement have been an indispensable help. With PhDArts she has established an extraordinary institute for exchange and reflection in the field of artistic research. Over the past twenty years, the conversations with co-supervisor Gert Staal about design, design education and how these fields are influenced by societal changes have been very valuable to me. The enthusiasm and versatile knowledge of co-promoter Lucas Evers, and the ability to point me towards new paths, has helped me enormously.

I was fortunate to do my research within the highly generous PhDArts program in which, together with a diverse group of researchers, I could develop and test my investigations. I am very grateful to my fellow candidates, the Royal Academy of Art The Hague and Leiden University, in particular Janneke Wesseling, director of PhDArts, Judith van IJken, Suzanne Knip-Mooij, Alice Twemlow, Judith Westerveld, Henk Borgdorff, academic director of the Academy of Creative and Performing Arts, and Rosalien van der Poel, its institute manager.

An important part of the research has been done within my design studio SJG. The visualizations in this book have been a joint effort by the entire studio for the past few years. Moreover, many of the questions for this research arose from the conversations in the studio. I thank Dimitri Jeannottat, Megan Adé, Julie da Silva, Salomé Bernhard, Clémence Guillemot, Silke Koeck, Yulia Kondratyeva, Raphael Mathias, Laura Opsomer, Simon Ruaut, Carina Schwake, Stella Shi, Linda Ursem and Denisse Vega de Santiago.

The themes investigated in this book are informed by discussions with students and tutors of the Information Design Master at the Design Academy Eindhoven. In addition to all of the students, I am grateful to Simon Davies, Gert Staal, Frans Bevers, Kim Bouvy, Toon Koehorst, Arthur Roeloffzen, Jantien Roozenburg, Irene Stracuzzi, Vincent Thornhill and Jannetje in 't Veld. I also thank the executive board of the Design Academy Eindhoven, my colleague Master heads Louise Schouwenberg and Jan Boelen, and the DAE Knowledge Circle.

For their help with the realization of this book, I am grateful to D'Laine Camp for her careful copy-editing, Keonaona Peterson for her thorough proofreading, Lars Müller and Maya Rüegg for their respectful guidance of this book towards a readership, the Creative Industries Fund NL for their generous support, and printer NPN and bookbinder Patist for their efforts to achieve perfection.

I owe a lot of thanks to my family for their patience, support and encouragement. Especially to Baukje Trenning, with whom I share so many things. I dedicate this book to her, and to Step, and Claar.

Imprint

Doctoral thesis by Joost Grootens, at PhDArts, Academy of Creative and Performing Arts, Leiden University and University of the Arts The Hague. Supervisor: Prof. dr. Janneke Wesseling. Co-supervisors: Lucas Evers and Gert Staal.

Visualizations	SJG / Megan Adé, Salomé Bernhard, Julie da Silva, Joost Grootens, Clémence Guillemot, Dimitri Jeannottat, Silke Koeck, Yulia Kondratyeva, Raphael Mathias, Laura Opsomer, Simon Ruaut, Carina Schwake, Stella Shi, Linda Ursem, Denisse Vega de Santiago.
Copy editing	D'Laine Camp
Proofreading	Keonaona Peterson
Book design	SJG / Joost Grootens, Dimitri Jeannottat
Printing	NPN Drukkers
Binding	Boekbinderij Patist

© 2021 Lars Müller Publishers, Zurich, and Joost Grootens, Amsterdam

Lars Müller Publishers is supported by the Swiss Federal Office of Culture with a structural contribution for the years 2016–2020.

Lars Müller Publishers
Zurich, Switzerland
www.lars-mueller-publishers.com

ISBN 978-3-03778-658-1

Distributed in North America by Artbook I D.A.P.
www.artbook.com

Printed in the Netherlands

creative industries
fund NL

This book was made possible by the generous support of the Creative Industries Fund NL.